JN302666

パイロットのための
ICAO航空英語能力試験ワークブック

Simon Cookson & Michael Kelly

READY FOR DEPARTURE！

Workbook
with SPRC model！

成山堂書店

本書の内容の一部あるいは全部を無断で電子化を含む複写複製（コピー）及び他書への転載は，法律で認められた場合を除いて著作権者及び出版社の権利の侵害となります。成山堂書店は著作権者から上記に係る権利の管理について委託を受けていますので，その場合はあらかじめ成山堂書店（03-3357-5861）に許諾を求めてください。なお，代行業者等の第三者による電子データ化及び電子書籍化は，いかなる場合も認められません。

Acknowledgments

A lot of people have contributed to this book, and we would like to especially thank:

- Mari Chibusa, Takuma Inukai, Ryota Ishitobi, Genki Kayama, Hironori Konagamitsu and Kazunori Munakata for testing the ideas and activities;
- Marina Kuribayashi, Ayumi Inoue and Suvd Jargalsaikhan for contributing all the pictures;
- Associate Professor Chihiro Tajima of Keisen University for advising on material design and helping with the Japanese sections;
- Akiko Inada of J. F. Oberlin University for writing the Japanese introduction;
- and Yuki Nakamura, Ryoki Otsu, Ryota Ishitobi and Iori Kuroda for recording the audio tracks.

While a lot of people have contributed to the book, we would like to make clear that any mistakes that remain in the manuscript are the sole responsibility of the authors.

This book could not have been published without the guidance and support of the staff of Seizando-Shoten Publishing Co. We are very grateful to Yosuke Itagaki, Takashi Ono and the company president, Noriko Ogawa, for all their patience!

To our readers, we wish you happy and safe flying!

Simon Cookson and Michael Kelly, August 2013

Flight Operations Program
Aviation Management Department
J. F. Oberlin University
Tokyo, Japan

CONTENTS

Section	Pages	Content of Section	Flight Phase
Introduction	2-5	The ICAO English Language Proficiency Test & the SPRC Model	
Unit 1	6-11	*"Ladies And Gentlemen…"* Delays Due To Weather	Pre-Flight Operations
Unit 2	12-17	*"Use Caution For Icy Conditions"* Snow & Ice At The Airport	At The Ramp
Unit 3	18-23	*"Hold Your Position!"* Obstructions On The Taxiway	Ground Movement
Unit 4	24-29	*"Stand By For Clearance"* Rejecting Takeoff	Cleared For Takeoff
Unit 5	30-35	*"Check And Confirm"* Air Turnback	Takeoff & Climb
Unit 6	36-41	*"Level Off At Your Discretion"* Equipment Failure & Rough Air	Climb
Unit 7	42-47	*"Is There A Doctor On Board?"* Passenger Injuries & Problems	Cruise
Unit 8	48-53	*"Mayday, Mayday, Mayday!"* Smoke In The Cabin	Emergency
Unit 9	54-59	*"The Airport Is Now Closed"* Bad Weather & Natural Disasters	Holding
Unit 10	60-65	*"Down And Locked?"* Problems During Approach	Approach
Unit 11	66-71	*"Go Around, Go Around"* Crosswinds & Wake Turbulence	Landing
Unit 12	72-77	*"Request Emergency Assistance"* Overruns & Other Mishaps	After Landing
Answer Key	78-89	Answers to unit activities	
Self-Evaluation	90	Checklist for the 6 ICAO Areas Of Evaluation	

INTRODUCTION

WORKBOOK CONTENTS

This workbook prepares you for the ICAO English language proficiency test and is designed for use <u>after</u> the *Ready for Departure* textbook. The workbook includes many activities that practice all 6 evaluation areas of the ICAO test. The activities help to improve your English through three important steps:
1. understanding key vocabulary
2. making sentences using key vocabulary
3. organizing your speech using sentences with key vocabulary

The workbook has 12 units with the same topics as the *Ready for Departure* textbook. Each unit has 6 pages:
- *Vocabulary:* reviews key words and their meanings for the unit;
- *Single Picture:* introduces the **SPRC model** to help organize your responses;
- *Useful Language:* extra exercises to practice grammar structures from *Ready for Departure*;
- *Picture Sequence:* introduces basic questions for the picture sequence section of the ICAO test;
- *Vocabulary Review:* extra exercises to practice making sentences with key words;
- *Single Picture Review:* an example **SPRC model** to improve your vocabulary range and fluency.

The units provide a lot of additional practice in all 6 evaluation areas: pronunciation, structure, vocabulary, fluency, comprehension and interactions. They also include useful advice to help improve your English proficiency and prepare you for the ICAO test.

Units 1-3 and 7-9 have a special focus on pronunciation, with exercises for the B/V, S/TH and L/R sounds, which are problematic for many Japanese people. Units 4-6 and 10-12 have extra comprehension exercises.

KEY POINTS TO REMEMBER

- *Use complete sentences!*
  It is important to control your sentence structure. Learn to master basic sentence structures, and practice speaking in complete sentences. Then your responses will be more accurate and easier to understand. When you gain more confidence, try to use more complex sentence structures.

- *Organize your responses!*
  Practice using the **SPRC model** to organize your responses. Your responses will be better if they are well organized. Practice finding the key elements in each picture with the help of the **SPRC model**.

- *Control your tenses!*
  Remember to control your tenses when describing the single picture and picture sequence. You must use the <u>present tense</u> for the single picture, but for the picture sequence you must use the <u>past tense</u>. The questions in the picture sequence exercises will help you stay in the past tense.

- *Make step-by-step improvements!*
  You need to score a minimum Level 4 in each of the 6 evaluation areas. Understand your strong and weak points, and work especially on your weak points. Use the Self-Evaluation checklist on page 90 to check your proficiency in each evaluation area.

- *Practice using your English!*
  You have to practice your English to get better. It takes time and hard work to improve your English proficiency, so don't wait to study! Find a friend and practice together now!

INTRODUCTION

WORKBOOK CONTENTS

本ワークブックは、ICAO 英語能力テスト準備用として作成されており、「パイロットのためのICAO航空英語能力試験教本」（以下：Ready for Departure）で学習した後に使用されることを前提としています。本書は数多くのアクティビティーを含んでおり、ICAOテストの6つの評価項目を全て練習していきます。アクティビティーでは、次の3つの重要なステップを通して英語力向上を図ります。
1. 重要な語彙の理解
2. 重要語彙を使った文章作成
3. 重要語彙を含んだ文章を使った回答の組み立て

本書は12ユニットで構成されており、Ready for Departure と同じトピックを使用しています。各ユニットは 6 ページで、以下を含みます：
- Vocabulary：各ユニットにおけるキーワードとその意味の復習
- Single Picture：回答の組み立てに役立つ SPRC model の紹介
- Useful Language：Ready for Departure で学習した文法構造の練習
- Picture Sequence：ICAO テストの Picture Sequence のための基本的な質問の紹介
- Vocabulary Review：キーワードを使った文章作成の練習
- Single Picture Review：語彙領域や流暢さに磨きをかける SPRC model の例

各ユニットでは、6 つの評価項目（発音、文法、語彙、流暢さ、理解、意思疎通）全てにおいて練習を追加しています。また、各ユニットには役立つアドバイスが入っており、英語力を向上させると共に ICAO テストの準備をしていきます。ユニット 1〜3 と 7〜9 は発音に特に焦点を当て、日本人の多くが問題としている B/V 音、S/TH 音、L/R 音を練習します。ユニット 4〜6 と 10〜12 には理解力訓練が入っています。

KEY POINTS TO REMEMBER

- *Use complete sentences!* 完全な文章を使いましょう！
  文章構造をコントロールすることが重要です。基本的な文章構造を学習し、完全な文章で話す練習をしましょう。そうすることで、回答はより正確で理解されやすくなります。自信が持てたら、より複雑な文章構造を使ってみましょう。

- *Organize your responses!* 回答を整理しましょう！
  SPRC modelを使って回答を整理する練習をしましょう。まとまりが良く整っていれば、回答はより良いものになります。**SPRC model**の助けを借りて、各絵の中にある重要な要素を見つけて練習しましょう。

- *Control your tenses!* 時制をコントロールしましょう！
  Single PictureやPicture Sequenceを描写するときは、時制に気をつけましょう。Single Pictureを描写する時は<u>現在形</u>を使いますが、Picture Sequenceには<u>過去形</u>を使います。Picture Sequenceの質問では、過去形を使う練習をします。

- *Make step-by-step improvements!* 着実な進歩を遂げましょう！
  6つの評価項目の各項目において、レベル4以上のスコアが必要になります。強みと弱点を理解し、特に弱点を中心に勉強しましょう。90ページにある自己評価チェックリストを使って各項目の英語力をチェックしましょう。

- *Practice using your English!* 英語を使う練習をしましょう！
  英語の上達には練習が必要です。英語力をつけるには時間と努力が必要ですから、待っていないで勉強を始めましょう！
  友達を見つけて今すぐ一緒に練習しましょう！

Practice is the best way to improve your English!

INTRODUCTION

THE 'SPRC' MODEL

One of the biggest problems for people taking the ICAO test is that responses are not well organized. If your responses are not well organized, they become difficult to understand and the meaning may be confused. On the other hand, if your responses are well organized, they are easier to understand and your meaning is clear. When a pilot communicates with the ATC controller, the messages must be short, simple and well organized.

We recognized this problem and found a way to make responses better organized. This is the **SPRC model**. **SPRC** stands for **S**ituation, **P**roblem, **R**esponse and **C**onclusion. This model helps you to see the important elements in a picture and organize them into responses that are easy to understand. Below is an example:

SITUATION
Understand the situation! In the picture you can see an airplane. The airplane is on a taxiway. You know it is a taxiway because there is a solid centerline. What is the airplane doing? The airplane is holding its position.

PROBLEM
Identify the problem! Why is the airplane holding? The reason is simple. There is an obstruction on the taxiway. It seems there is some kind of fluid on the taxiway. That is the problem.

RESPONSE
Describe the response! How is the pilot responding to this problem? Now he is holding his position. Next he will contact the controller. Then the controller will send airport staff to the taxiway to remove the obstruction.

CONCLUSION
Check the conclusion! What is the result of the response? Airport staff will remove the obstruction. Then the taxiway will be safe for operations. Finally the pilot will be able to continue taxi to the runway.

The mindmap on the right shows the main elements of the picture. In each unit of this workbook there is a mindmap on the single picture page. The mindmaps help you to organize your responses. In addition, the questions on the picture sequence pages also help you to practice using the **SPRC model**.

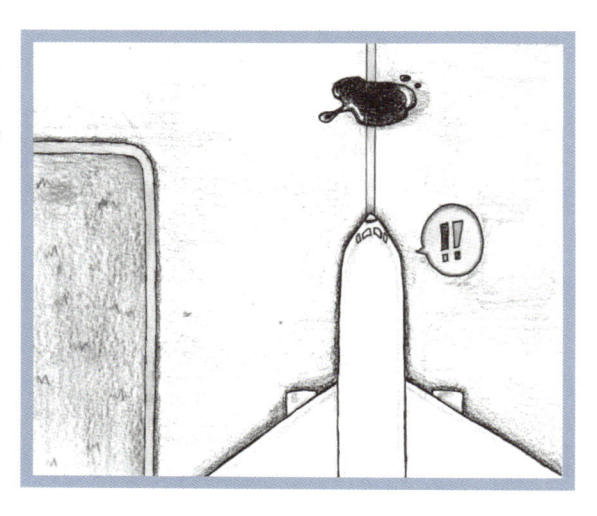

Use the SPRC model to organize your responses!

SITUATION
- airplane on taxiway
- hold position

⇩

PROBLEM
- obstruction
- some kind of fluid

⇩

RESPONSE
- contact controller
- remove obstruction

⇩

CONCLUSION
- safe for operations
- continue taxi

INTRODUCTION

THE 'SPRC' MODEL

ICAOテストで最も問題になることは、回答者の回答が整理されていないことです。回答がまとまっていないと理解されにくく、意味合いに混乱が生じます。一方、回答が良く整理されていれば理解されやすく、意味もはっきり伝わります。パイロットが航空管制官と意思疎通を図るとき、メッセージは短く簡単で、よく整理されたものでなければなりません。私たちはこの問題を認識し、より整理された回答ができる手段を見つけました。それが、**SPRC model**です。SPRCとは、状況(**S**ituation)、問題(**P**roblem)、対応(**R**esponse)、結論(**C**onclusion)の略です。このモデルでは、各絵の中にある重要な要素を見つけ出し、分かりやすい回答に整理する手がかりにします。以下は一例です：

SITUATION
Understand the situation! 状況を理解しましょう！ 絵には航空機が見えます。航空機は空港の誘導路にいます。実線のセンターラインがあるので、それが誘導路だと分かります。航空機は何をしているのでしょう。航空機はその場に留まっています。

PROBLEM
Identify the problem! 問題を認識しましょう！ なぜ航空機はその場に留まっているのでしょう。答えは簡単です。誘導路に障害物があるからです。誘導路に何らかの液体があるようです。それが問題なのです。

RESPONSE
Describe the response! 対応を描写しましょう！ パイロットはこの問題にどう対応するでしょう。今、彼はその場で待機しています。次に管制官に連絡するでしょう。そして管制官は障害物を取り除くために空港スタッフを誘導路に送り出すでしょう。

CONCLUSION
Check the conclusion! 結論を確認しましょう！ 対応の結果は何でしょう。空港スタッフは障害物を取り除くでしょう。そして誘導路の安全は確保されるでしょう。最終的に、パイロットは誘導路で地上走行を続けられるでしょう。

右にあるマインドマップは、絵の主な要素を表します。本書の各ユニットのSingle Pictureのページには、マインドマップがあります。マインドマップは回答の整理に役立ちます。さらに、Picture Sequenceのページにある質問は、SPRC modelを使う練習に効果的です。

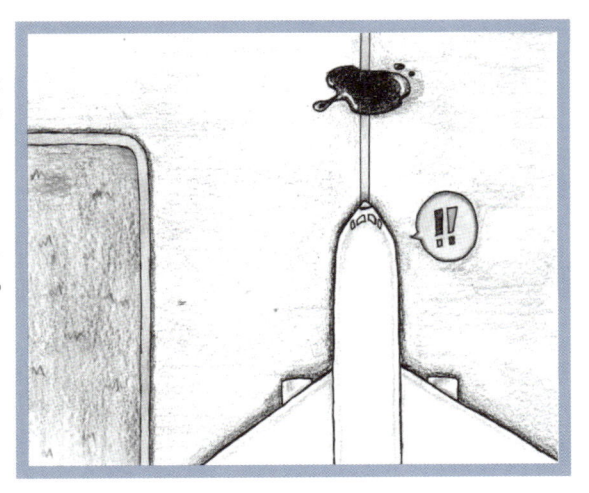

Good luck! Do your best! Get ready for departure!

SITUATION
- airplane on taxiway
- hold position

⇩

PROBLEM
- obstruction
- some kind of fluid

⇩

RESPONSE
- contact controller
- remove obstruction

⇩

CONCLUSION
- safe for operations
- continue taxi

UNIT 1   PRE-FLIGHT OPERATIONS

"LADIES AND GENTLEMEN..."
DELAYS DUE TO WEATHER

VOCABULARY

KEY WORDS

それぞれのキーワードにあった意味を選びましょう。例を参考にしてください。（回答は 78 ページ）

1. adverse weather   _h_   a. a period of time by which a flight is late or postponed
2. delay   b. something dangerous that may cause damage to aircraft or people
3. gate   c. a flash of bright light in the sky caused by an electrical discharge
4. hazard   d. a very powerful and sudden downward burst of wind
5. holding pattern   e. an electrical storm with lightning, heavy rain and strong gusting winds
6. lightning   f. a long level piece of ground used for takeoffs and landings
7. microburst   g. sudden and sometimes violent movement of air
8. runway   h. bad weather conditions that may restrict flight operations
9. thunderstorm   i. a circular or oval flight path used by aircraft waiting to land
10. turbulence   j. part of an airport terminal where passengers board or leave aircraft

PRONUNCIATION   CD1

CDで上の文を聞き、後についてリピートしましょう。何回か繰り返します。

PRONUNCIATION   CD2

CDで以下の表現を聞き、後についてリピートしましょう。

"B" IN INITIAL POSITION

1. **b**earing
2. **b**lind spot
3. **b**irds
4. **b**oarding **b**ridge
5. airport **b**eacon
6. **b**roken
7. **b**last fence
8. hot **b**rakes
9. lost **b**aggage
10. weight and **b**alance

To make the "B" sound: put your lips together, then open them quickly and use your voice!

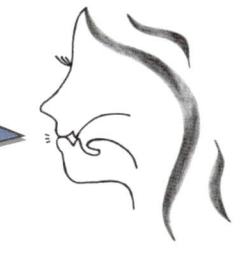

PRONUNCIATION   CD3

CDで以下の表現を聞き、後についてリピートしましょう。

"V" IN INITIAL POSITION

1. **v**ector
2. **v**ery humid
3. **v**elocity
4. **v**isual flight rules
5. magnetic **v**ariation
6. **v**icinity
7. fuel **v**apor
8. **v**olcano
9. ground **v**isibility
10. **v**ertical separation

To make the "V" sound: put your top front teeth and bottom lip together, then open your mouth and use your voice!

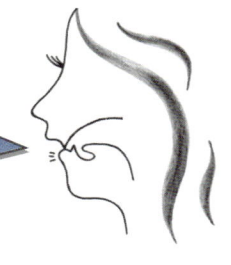

UNIT 1 PRE-FLIGHT OPERATIONS

SINGLE PICTURE

FLUENCY

絵の状況について、SPRC モデルを使って、3 分程度話しましょう。まず、左下の文を使って話し始めます。次に、右下のマインドマップの言葉を使って話すようにします。それを S、P、R、C とそれぞれ行います。

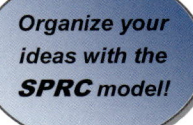

Organize your ideas with the SPRC model!

S = SITUATION

In this picture a severe thunderstorm is passing over an airport…

Thunderstorm	Runway & taxiways	Holding patterns
• where? • sky condition	• any aircraft? • wind condition	• where? • how many aircraft?

P = PROBLEM

Thunderstorms can be very dangerous and they can produce several different hazards…

Hazard 1	Hazard 2	Hazard 3
• lightning strike • what damage may happen?	• heavy rain • visibility • slippery	• strong winds • turbulence • microbursts

R = RESPONSE

I think that this airport is now closed because of the adverse weather…

Airport closed	Holding patterns
• why? • departure delays • passengers at gates	• what options? • landing delays • divert to alternate

C = CONCLUSION

After the thunderstorm passes the airport, I think that the weather will improve…

After thunderstorm
• weather improve
• airport re-open

Use present tenses for the single picture!

7

UNIT 1　　　　　　　　　　　　　　　　　　　　PRE-FLIGHT OPERATIONS

USEFUL LANGUAGE

STRUCTURE

このエクササイズでは、助動詞の「may」「might」「could」を使って、未来に起こりえる問題を表わします。下の単語を並べ替えて文を作りましょう。1問目の例を参考にしてください。

POSSIBLE PROBLEMS IN THE FUTURE

1. visibility / Heavy rain / reduce / for pilots. / may
   → *Heavy rain may reduce visibility for pilots.*
2. might / slippery. / make / the runways / Rain
   → _____
3. runways / cause / during landing. / could / Slippery / problems
   → _____

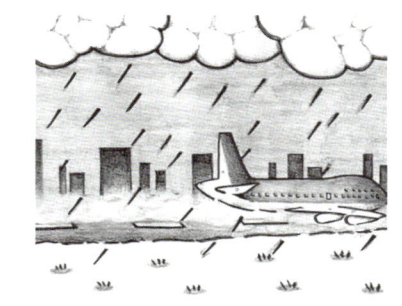

4. Heavy / flight delays. / cause / might / snow
   → _____
5. cause / aircraft wings. / icing / may / Snow / on
   → _____
6. stall. / could / on the wings / make / Icing / the aircraft
   → _____

7. winds / create / Strong / turbulence. / may
   → _____
8. injuries / Turbulence / cause / passengers. / to / could
   → _____
9. control / might / severe turbulence. / during / Pilots / lose
   → _____

10. a lot of / may / thunder and lightning. / produce / A thunderstorm
    → _____
11. some / Loud thunder / passengers. / scare / could
    → _____
12. systems. / might / an aircraft's / Lightning / electrical / damage
    → _____

PRONUNCIATION

CD4 CDで上の文を聞き、後についてリピートしましょう。何回か繰り返します。（回答は78ページ）

FLUENCY

上の4つの状況と同じような経験があれば、描写しながら話しましょう。

UNIT 1   PRE-FLIGHT OPERATIONS

PICTURE SEQUENCE

Use past tenses for the picture sequence!

INTERACTIONS

CD5 1～4の絵について以下の問題に答えましょう。まずは、CDでそれぞれの問題を聞きます。問題を聞いた後に一時停止ボタンを押して、それぞれの問題に対して回答します。完全文を使って、回答は声に出して言いましょう。

Picture 1

1. What was the situation at this airport?
2. What kind of problems did the weather cause?

Picture 2

3. How did the problems affect the passengers?
4. How did the passengers respond?

Picture 3

5. How did the situation change?
6. What did the controller do?

Picture 4

7. What happened after the weather changed?
8. Did the airport staff deal with this situation well?

UNIT 1   PRE-FLIGHT OPERATIONS

VOCABULARY REVIEW

VOCABULARY

キーワードの復習です。ヒントを読んで、クロスワードに答えを書きましょう。（回答は 78 ページ）

ACROSS

4. a circular or oval flight path used by aircraft waiting to land
7. part of an airport terminal where passengers board or leave aircraft
8. a long level piece of ground used for takeoffs and landings
9. sudden and sometimes violent movement of air
10. a very powerful and sudden downward burst of wind

DOWN

1. an electrical storm with lightning, heavy rain and strong gusting winds
2. bad weather conditions that may restrict flight operations
3. something dangerous that may cause damage to aircraft or people
5. a flash of bright light in the sky caused by an electrical discharge
6. a period of time by which a flight is late or postponed

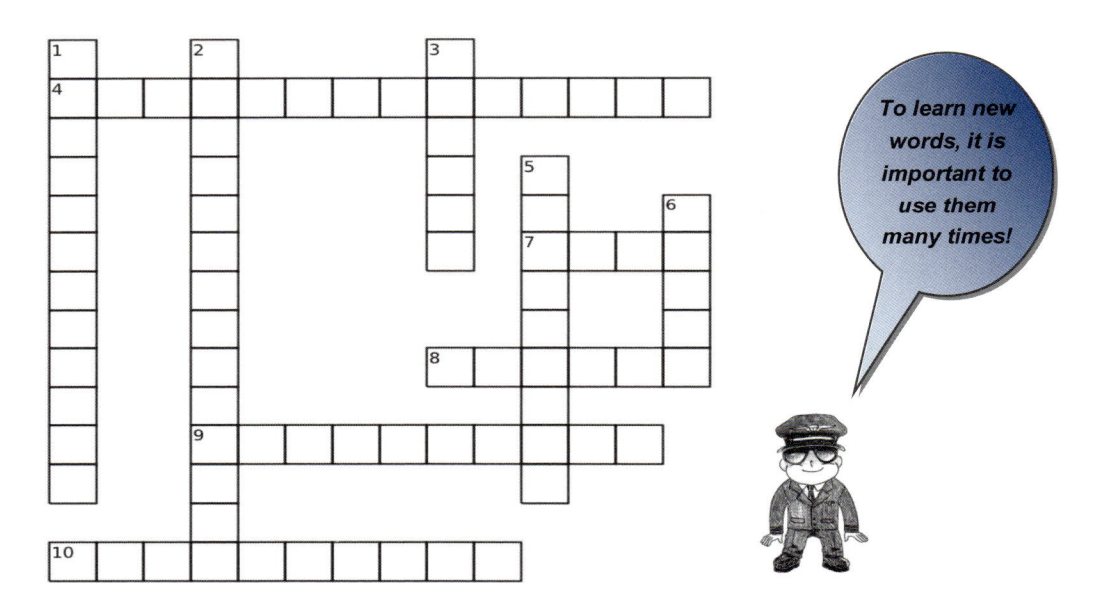

To learn new words, it is important to use them many times!

FLUENCY

上のいくつかのキーワードを使って、3 つの文を作りましょう。その後文を何回か音読します。

SENTENCES

1. _____

2. _____

3. _____

UNIT 1

PRE-FLIGHT OPERATIONS

SINGLE PICTURE REVIEW

VOCABULARY

以下のキーワードを使って、下線部を埋めましょう。(回答は 78 ページ)

KEY WORDS

a. adverse weather
b. delay
c. gates
d. hazards
e. holding patterns
f. lightning
g. microbursts
h. runway
i. thunderstorm
j. turbulence

SITUATION

In this picture a severe __1._____ is passing over an airport. The sky condition is dark, there is a lot of lightning, and I think that there is also heavy rain. I can see a wind indicator, and it looks like a strong crosswind is blowing across the runway. I can see no aircraft on the __2._____ or taxiways, but two aircraft are in holding patterns above the airport.

PROBLEM

Thunderstorms can be very dangerous and they can produce several different __3._____. First, lightning is dangerous for aircraft and for airport operations. If __4._____ strikes an aircraft or an airport building, it may damage electrical systems. Second, heavy rain or sleet may reduce visibility and make the runways slippery. Therefore takeoffs and landings might become dangerous. In addition, thunderstorms sometimes cause strong winds and turbulence. It can be difficult to fly in these conditions, and passengers could be injured by severe __5._____. Finally, there is a risk of microbursts. __6._____ are difficult to see and they can be very dangerous for aircraft on final approach.

RESPONSE

I think that this airport is now closed because of the __7._____. No aircraft are taking off or landing at the moment because the thunderstorm is over the airport and the crosswinds are too strong. Therefore aircraft departures will be delayed or cancelled, and I think that many passengers may have to wait at the __8._____. Also, two aircraft are in holding patterns above the airport. These airplanes are waiting for clearance to land, and their landings will be delayed. The pilots of these aircraft should think about their options. If the __9._____ is too long, the pilots may decide to divert to alternate airports.

CONCLUSION

After the thunderstorm passes the airport, I think that the weather will improve and the airport can re-open. Then the delayed aircraft will be able to take off. Also, the aircraft in __10._____ will receive clearance to land.

FLUENCY

CD6 CDで上の文を聞き、音読しましょう。何回か音読を繰り返します。

UNIT 2 — AT THE RAMP

"USE CAUTION FOR ICY CONDITIONS"

SNOW & ICE AT THE AIRPORT

VOCABULARY

それぞれのキーワードにあった意味を選びましょう。例を参考にしてください。（回答は 79 ページ）

KEY WORDS

1. boarding stairs — d — a. stopping or terminating a flight or ATC clearance
2. cancellation — ___ — b. a careful examination of an aircraft before departure
3. cargo — ___ — c. a path at an airport that connects runways with terminals and ramps
4. de-icing — ___ — d. steps used for getting on or off aircraft
5. fuselage — ___ — e. part of the airport where aircraft are parked, loaded and boarded
6. precaution — ___ — f. actions taken during cold weather conditions to promote flight safety
7. pre-flight inspection — ___ — g. the main body of an aircraft that holds crew, passengers and cargo
8. ramp — ___ — h. protection against some kind of danger
9. taxiway — ___ — i. removing ice or snow from an aircraft using fluids, heating or boots
10. winter operations — ___ — j. goods or products that are carried by aircraft or ships

PRONUNCIATION — CD7

CDで上の文を聞き、後についてリピートしましょう。何回か繰り返します。

PRONUNCIATION — CD8

CDで以下の表現を聞き、後についてリピートしましょう。

"S" IN INITIAL POSITION

1. **s**nowstorm
2. blowing **s**now
3. **s**leet
4. bird **s**ighting
5. warning **s**ignal
6. **s**ituation
7. runway **s**ign
8. **s**tall warning
9. weather **s**ervice
10. **s**quawk **s**even-**s**even-zero-zero

To make the "S" sound: put the tip of your tongue behind the bottom front teeth!

PRONUNCIATION — CD9

CDで以下の表現を聞き、後についてリピートしましょう。

"TH" IN INITIAL POSITION

1. **th**understorm
2. **th**in ice
3. **th**ird
4. **th**ick fog
5. spot **th**irteen
6. runway **th**reshold
7. **th**undershower
8. apply full **th**rottle
9. **th**ermal column
10. runway **th**ree six

To make the "TH" sound: put the tip of your tongue between the top and bottom front teeth!

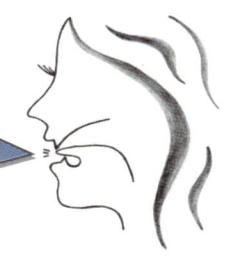

UNIT 2 — AT THE RAMP

SINGLE PICTURE

FLUENCY

絵の状況について、SPRC モデルを使って、3 分程度話しましょう。まず、左下の文を使って話し始めます。次に、右下のマインドマップの言葉を使って話すようにします。それを S、P、R、C とそれぞれ行います。

*Organize your ideas with the **SPRC** model!*

S = SITUATION

The weather is very bad at this airport…

- Weather
  - snow & ice
  - winter operations
- Hangar
  - any aircraft?
  - why?
- Ramp
  - any aircraft?
  - any vehicles?

P = PROBLEM

Snow and ice are hazards for flight operations…

- Hazards
  - ice on ramp, taxiways & runways
  - icing on wings
  - slippery boarding stairs

R = RESPONSE

The airport staff must do several things to prepare for winter operations…

- Snow removal
  - snowplow
  - ramp area
  - taxiways & runways
- Aircraft on ramp
  - de-icing
  - pre-flight inspection
- Boarding stairs
  - icy
  - clean & inspect

C = CONCLUSION

Safety is always important in flight operations…

- Safety
  - winter operations
  - precautions

Make sentences with the words in the mindmap!

UNIT 2 — AT THE RAMP

USEFUL LANGUAGE

STRUCTURE

このエクササイズでは、単純過去形を使って、過去に起きた事を表わします。下の単語を並べ替えて文を作りましょう。1問目の例を参考にしてください。

EVENTS IN THE PAST

1. were / and / One aircraft / on the ramp. / several vehicles
   → *One aircraft and several vehicles were on the ramp.*

2. the ramp / and ice. / cleared / of snow / Airport staff
   → _____

3. and fuselage / accumulated / the plane. / on the wings / Snow / of
   → _____

4. for / inspection. / went outside / The pilot / the pre-flight
   → _____

5. The pilot / a landing gear tire. / a problem / found / with
   → _____

6. asked / the tire. / to / maintenance staff / The pilot / fix
   → _____

7. the cargo / The ground crew / the aircraft. / on to / loaded
   → _____

8. the bags / and secured / A baggage handler / in the cargo hold. / loaded
   → _____

9. dangerous / checked / goods. / for / The security dog / the bags
   → _____

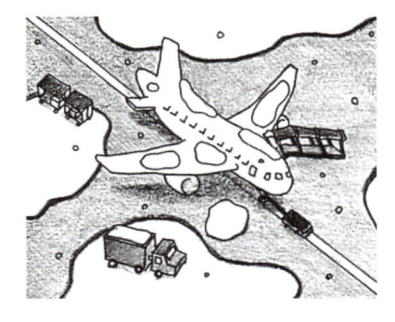

10. a baggage car / next to / A fuel truck / the plane. / were / and
    → _____

11. with fuel / was / the fuel truck. / loaded / The aircraft / from
    → _____

12. fuselage. / were / the aircraft's / from / The boarding stairs / removed
    → _____

PRONUNCIATION

CD10 CDで上の文を聞き、後についてリピートしましょう。何回か繰り返します。(回答は79ページ)

FLUENCY

上の4つの状況と同じような経験があれば、描写しながら話しましょう。

UNIT 2   AT THE RAMP

PICTURE SEQUENCE

Try to speak in complete sentences!

INTERACTIONS

CD11 1〜4の絵について以下の問題に答えましょう。まずは、CDでそれぞれの問題を聞きます。問題を聞いた後に一時停止ボタンを押して、それぞれの問題に対して回答します。完全文を使って、回答は声に出して言いましょう。

Picture 1

1. What was the situation at this airport?
2. What kind of problems did the weather cause?

Picture 2

3. What did the ground crew do to the airplane?
4. Why did they do this?

Picture 3

5. What did the ground crew do next?
6. What precautions did the ramp staff take?

Picture 4

7. Describe the problems on the taxiway.
8. What do you think happened next?

UNIT 2 — AT THE RAMP

VOCABULARY REVIEW

VOCABULARY

キーワードの復習です。ヒントを読んで、クロスワードに答えを書きましょう。（回答は 79 ページ）

ACROSS

2. steps used for getting on or off aircraft
5. part of the airport where aircraft are parked, loaded and boarded
7. actions taken during cold weather conditions to promote flight safety
9. stopping or terminating a flight or ATC clearance
10. goods or products that are carried by aircraft or ships

DOWN

1. a careful examination of an aircraft before departure
3. the main body of an aircraft that holds crew, passengers and cargo
4. removing ice or snow from an aircraft using fluids, heating or boots
6. protection against some kind of danger
8. a path at an airport that connects runways with terminals and ramps

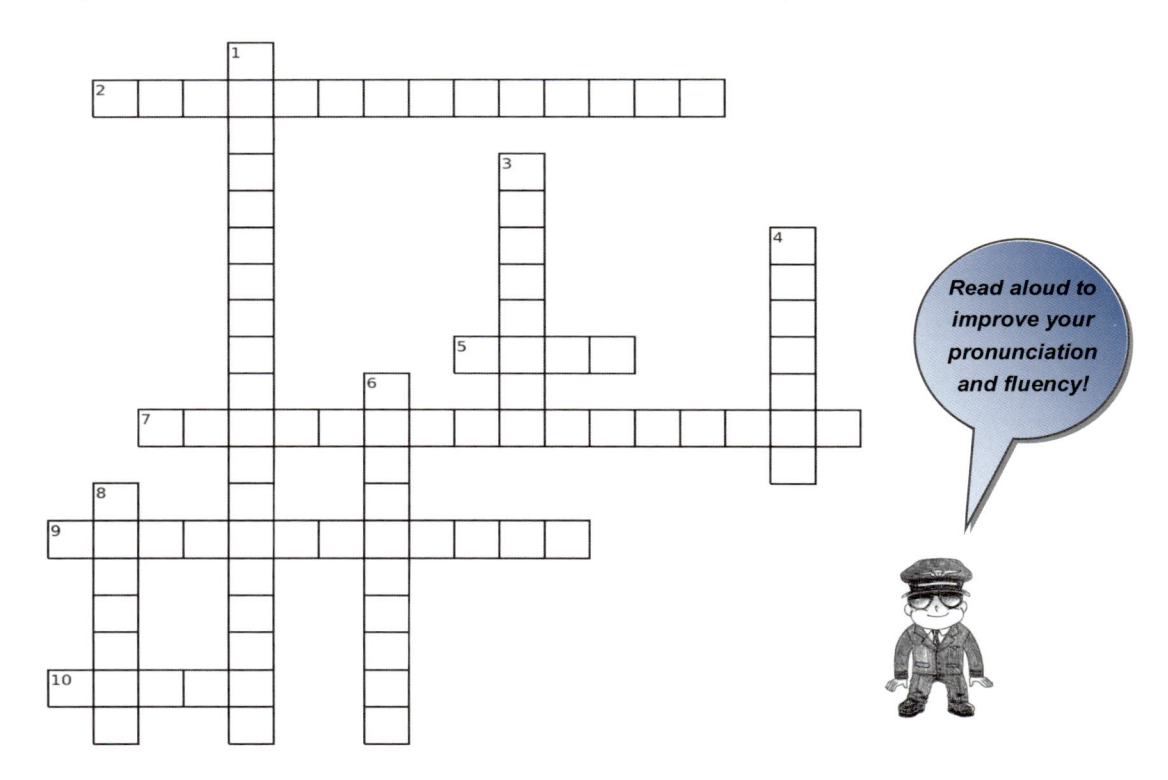

Read aloud to improve your pronunciation and fluency!

FLUENCY

上のいくつかのキーワードを使って、3 つの文を作りましょう。その後文を何回か音読します。

SENTENCES

1. _____
2. _____
3. _____

UNIT 2　　　　　　　　　　　　　　　　　　　　　　　　　　　　　AT THE RAMP

SINGLE PICTURE REVIEW

VOCABULARY

以下のキーワードを使って、下線部を埋めましょう。（回答は 79 ページ）

KEY WORDS

a. boarding stairs
b. cancellations
c. cargo
d. de-ice
e. fuselage
f. precautions
g. pre-flight inspection
h. ramp
i. taxiways
j. winter operations

SITUATION

The weather is very bad at this airport. It is snowing, and there is snow and ice all around the __1._____ area. It looks like winter operations are in effect at the airport. I can see a large hangar. One of the hangar doors is open, and there is one aircraft inside. Aircraft are sometimes moved to the hangars when the weather is very bad. There is another airplane on the ramp in front of the hangar. This plane has a lot of ice and snow on the wings and __2._____. I can also see a snowplow and a portable boarding stairs vehicle on the ramp. __3._____ vehicles are used for open spots at the airport.

PROBLEM

Snow and ice are hazards for flight operations. Sometimes they cause flight delays or __4._____. Ice is slippery and so it can make the ramp area, __5._____ and runways dangerous. Also, icing on the wings of an aircraft is very dangerous because it might cause the aircraft to stall after takeoff. Finally, if there is ice on the boarding stairs then passengers may slip and injure themselves.

RESPONSE

The airport staff must do several things to prepare for __6._____. First, in the picture I can see that the snowplow is removing snow from the ramp area. In addition, the ground staff must remove snow from the taxiways and the runways to make them safe. Second, they need to __7._____ the airplane on the ramp before it can takeoff. Also, the ground staff must carefully load the __8._____ and fuel on to the plane. The flight crew have to do a __9._____ to make sure that the aircraft is ready for flight. Finally, the boarding stairs might be very icy so the ground staff should carefully clean and inspect them. Then the passengers can board the airplane.

CONCLUSION

Safety is always important in flight operations, but it is especially important during winter operations. If the ground staff and flight crews take __10._____, then safe flight operations can continue.

FLUENCY

CD12 CDで上の文を聞き、音読しましょう。何回か音読を繰り返します。

UNIT 3 — GROUND MOVEMENT

"HOLD YOUR POSITION!"
OBSTRUCTIONS ON THE TAXIWAY

VOCABULARY

それぞれのキーワードにあった意味を選びましょう。例を参考にしてください。(回答は 80 ページ)

KEY WORDS

1. baggage — b
2. baggage car — ___
3. braking action — ___
4. hold your position — ___
5. hydraulic fluid — ___
6. leak — ___
7. obstruction — ___
8. pushback — ___
9. slippery — ___
10. taxi — ___

a. using a ground vehicle to move an aircraft from the gate to a taxiway
b. bags and suitcases used by passengers for travel, also called luggage
c. an instruction given by a controller for a pilot to stop movement
d. an object that blocks the movement of an aircraft or airport vehicle
e. a condition where people or vehicles may lose control of movement
f. how easily an aircraft can stop on a runway during takeoff or landing
g. a liquid used to transfer power in aircraft systems (eg: brakes, steering)
h. a ramp vehicle used to transport checked-in bags and suitcases
i. slow movement of a plane on the ground before takeoff or after landing
j. a liquid or gas escaping from a hole in a pipe or container

PRONUNCIATION

CD13 CDで上の文を聞き、後についてリピートしましょう。何回か繰り返します。

PRONUNCIATION

CD14 CDで以下の表現を聞き、後についてリピートしましょう。

"L" IN INITIAL POSITION

1. light signals
2. localizer
3. landing sequence
4. boundary lights
5. lost communications
6. forced landing
7. enter left base
8. loud and clear
9. clearance limit
10. lateral separation

To make the "L" sound: touch the tip of your tongue to the top of your mouth!

PRONUNCIATION

CD15 CDで以下の表現を聞き、後についてリピートしましょう。

"R" IN INITIAL POSITION

1. readback
2. readout
3. radio station
4. release
5. runway three three right
6. enter right base
7. flight recorder
8. reporting point
9. fuel remaining
10. airport surveillance radar

To make the "R" sound: lift the middle of your tongue but don't touch the top of your mouth!

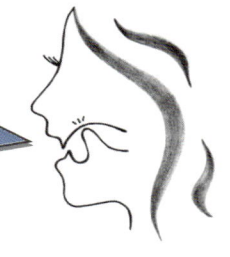

UNIT 3 — GROUND MOVEMENT

SINGLE PICTURE

FLUENCY

絵の状況について、SPRC モデルを使って、3 分程度話しましょう。まず、左下の文を使って話し始めます。次に、右下のマインドマップの言葉を使って話すようにします。それを S、P、R、C とそれぞれ行います。

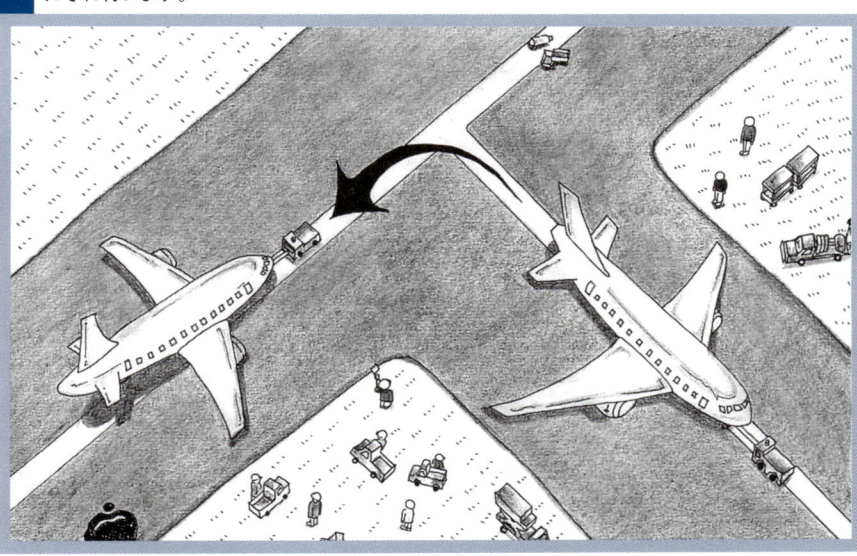

*Organize your ideas with the **SPRC** model!*

S = SITUATION

An airplane is being pushed back from the gate area…

Airplane	Ground vehicles
• pushback • from gate to taxiway	• baggage car • fuel truck • pushback car

P = PROBLEM

However, I can see some obstructions on the taxiway…

Obstruction 1	Obstruction 2
• trash • bottle & boxes • center of taxiway	• oil or hydraulic fluid • slippery • braking action

R = RESPONSE

There are many workers in this area, but they have not noticed the obstructions…

Remove obstructions
- report to controller
- remove trash
- clean up oil or hydraulic fluid

C = CONCLUSION

After the ground staff remove the obstructions, I think the pilot can taxi to the runway…

After removal
- taxi to runway
- check for obstructions

Use present tenses for the single picture!

UNIT 3   GROUND MOVEMENT

USEFUL LANGUAGE

STRUCTURE

このエクササイズでは、「some kind of 」「seems to be」「looks like」の表現を使って、障害物を表わします。下の単語を並べ替えて文を作りましょう。1問目の例を参考にしてください。

DESCRIBING OBSTRUCTIONS

1. obstruction / There is / the airplane. / in front of / some kind of
   → *There is some kind of obstruction in front of the airplane.*
2. and some boxes / looks like / It / on the taxiway. / a bottle / are
   → _____
3. contact / the obstruction. / to remove / will / the controller / The pilot
   → _____
4. many / the ramp area. / There / in / are / vehicles 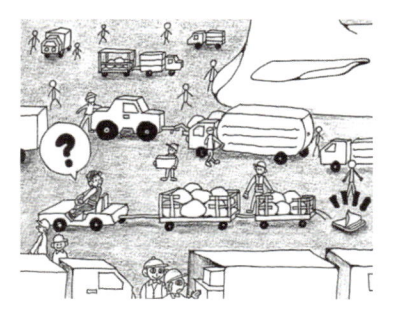
   → _____
5. looks like / from / It / something / the baggage car. / fell
   → _____
6. I think / the problem. / that / of the baggage car / the driver / noticed
   → _____
7. an obstruction / the taxiway. / to be / on / There seems 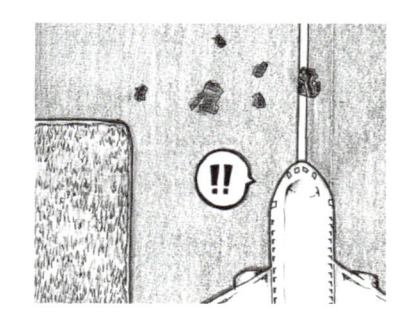
   → _____
8. some / and clothes. / It / work gloves / looks like
   → _____
9. because / the airplane / The pilot / the obstruction. / of / stopped
   → _____
10. looks like / It / the airplane. / fell from / something 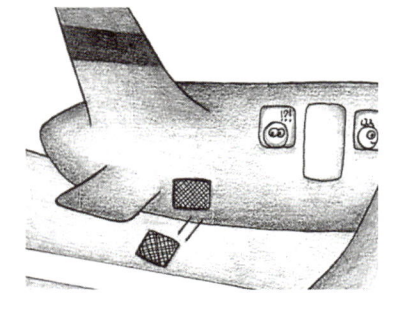
    → _____
11. the fuselage. / a panel / seems / from / to be / It
    → _____
12. will / the missing panel. / about / contact / The controller / the pilot
    → _____

PRONUNCIATION

CD16 CDで上の文を聞き、後についてリピートしましょう。何回か繰り返します。(回答は80ページ)

FLUENCY

上の4つの状況と同じような経験があれば、描写しながら話しましょう。

UNIT 3   GROUND MOVEMENT

PICTURE SEQUENCE

Use past tenses for the picture sequence!

INTERACTIONS   **CD17** 1～4の絵について以下の問題に答えましょう。まずは、CDでそれぞれの問題を聞きます。問題を聞いた後に一時停止ボタンを押して、それぞれの問題に対して回答します。完全文を使って、回答は声に出して言いましょう。

Picture 1
1. What was the situation at this airport?
2. What were the ramp workers doing?

Picture 2
3. What was the problem?
4. Did the ground staff notice the problem?

Picture 3
5. How did the pilots respond to the problem?
6. What did the controller do?

Picture 4
7. What did the ground staff do?
8. What do you think happened next?

UNIT 3　　　　　　　　　　　　　　　　　　　　　　　GROUND MOVEMENT

VOCABULARY REVIEW

VOCABULARY

キーワードの復習です。ヒントを読んで、クロスワードに答えを書きましょう。(回答は 80 ページ)

ACROSS

2. an object that blocks the movement of an aircraft or airport vehicle
7. using a ground vehicle to move an aircraft from the gate to a taxiway
8. bags and suitcases used by passengers for travel, also called luggage
9. a ramp vehicle used to transport checked-in bags and suitcases
10. a liquid used to transfer power in aircraft systems (eg: brakes, steering)

DOWN

1. an instruction given by a controller for a pilot to stop movement
3. slow movement of a plane on the ground before takeoff or after landing
4. a condition where people or vehicles may lose control of movement
5. how easily an aircraft can stop on a runway during takeoff or landing
6. a liquid or gas escaping from a hole in a pipe or container

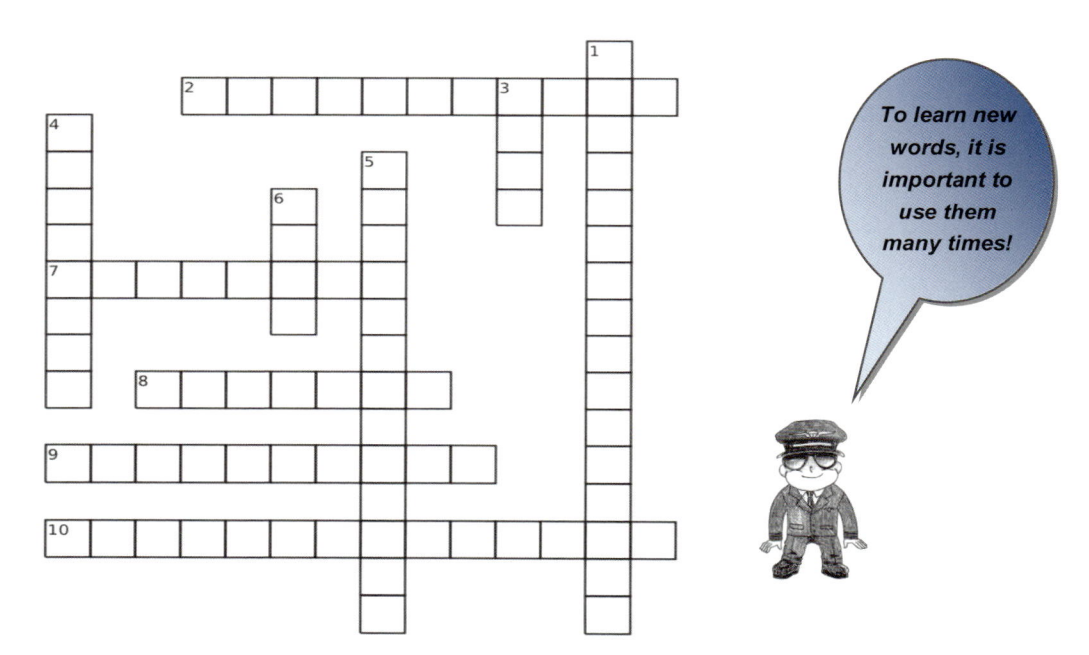

To learn new words, it is important to use them many times!

FLUENCY

SENTENCES

上のいくつかのキーワードを使って、3 つの文を作りましょう。その後文を何回か音読します。

1. _____

2. _____

3. _____

UNIT 3 GROUND MOVEMENT

SINGLE PICTURE REVIEW

VOCABULARY

以下のキーワードを使って、下線部を埋めましょう。（回答は 80 ページ）

KEY WORDS

a. baggage
b. baggage car
c. braking action
d. hold his position
e. hydraulic fluid
f. leaks
g. obstructions
h. pushback
i. slippery
j. taxi

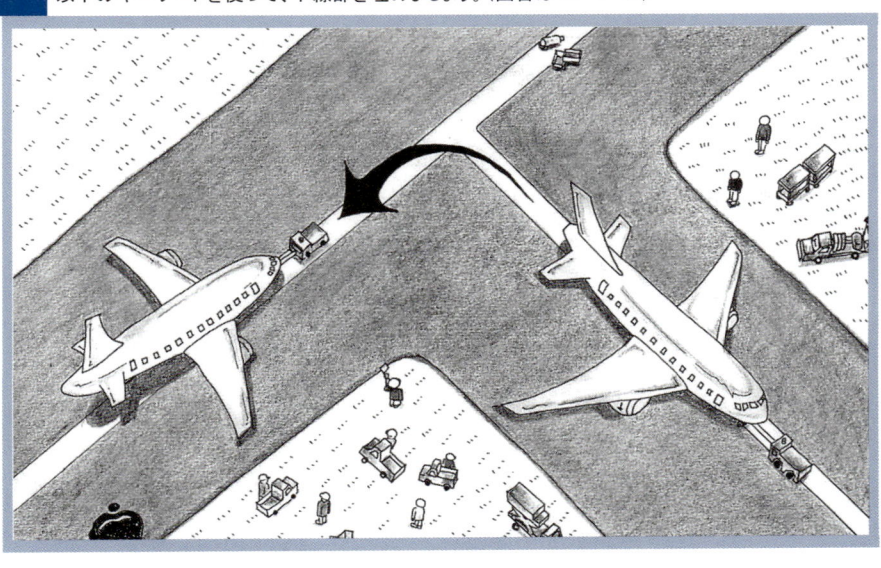

SITUATION

An airplane is being pushed back from the gate area. I can see many vehicles around the plane. There is a ___1.___ that is used to carry the passengers' ___2.___. I think that there is also a fuel truck and a vehicle that carries food and drinks. A pushback car is attached to the nose of the airplane. The driver of the ___3.___ car is moving the aircraft onto the taxiway.

PROBLEM

However, I can see some ___4.___ on the taxiway. There seems to be some trash in front of the airplane. It looks like a large bottle and some boxes. This obstruction is near the center of the taxiway. It is dangerous because even small objects can cause damage to a plane. I can also see some kind of oil or hydraulic fluid on the taxiway. Sometimes oil or hydraulic fluid ___5.___ from aircraft onto taxiways or runways. This can make the surface very slippery and it can affect the ___6.___ of airplanes.

RESPONSE

There are many workers in this area, but they have not noticed the obstructions. Someone must report the obstructions. Then, the controller can order the ground staff to remove them from the taxiway. Now the pilot must stop taxiing because of the obstruction. The pilot will have to ___7.___ until someone picks up the objects from the taxiway. I think that this should not cause a long delay. It is easy to remove obstructions like bottles or boxes, but it is more difficult to clean up oil and ___8.___. It takes time for workers to completely remove the fluids so that the surface is not ___9.___.

CONCLUSION

After the ground staff remove the obstructions, I think the pilot can ___10.___ to the runway. Safety is very important during operations on the ground. Ramp workers must always check and confirm that the area around the airplane is clear of obstructions. The ramp crew need good coordination to ensure the highest level of safety at the ramp and during ground movement.

FLUENCY

CD18 CDで上の文を聞き、音読しましょう。何回か音読を繰り返します。

UNIT 4   CLEARED FOR TAKEOFF

"STAND BY FOR CLEARANCE"
REJECTING TAKEOFF

VOCABULARY

それぞれのキーワードにあった意味を選びましょう。例を参考にしてください。（回答は 81 ページ）

KEY WORDS

1. bird strike   c
2. bird sweep   ___
3. fire equipment   ___
4. go around   ___
5. incident   ___
6. landing gear   ___
7. rejected takeoff   ___
8. re-schedule   ___
9. stand by   ___
10. tow assistance   ___

a. a controller or pilot must wait for a short time
b. the structure that supports an aircraft on the ground and allows it to taxi
c. a collision between a plane and a bird, usually during takeoff or landing
d. to change an event to a later time
e. an aborted landing for an aircraft on final approach
f. a procedure to remove birds from an area of the airport
g. using a ground vehicle to move an aircraft that has a problem
h. an event that happens during operations which affects aircraft safety
i. things that are used to put out a fire (eg: fire extinguisher, foam)
j. a situation in which a pilot decides to abort takeoff

PRONUNCIATION CD19
CDで上の文を聞き、後についてリピートしましょう。何回か繰り返します。

COMPREHENSION CD20
CDで 5 つのATCの会話を聞き、下線部を埋めましょう。（回答は 81 ページ）

LISTENING TO ATC

1. CONTROLLER: Citation 1663, cancel takeoff clearance, _____. There seems to be an obstruction on the runway near _____ intersection.
2. PILOT: TWR, Citation 1663, holding on _____, we see birds at the end of the runway. Request _____ before takeoff.
3. PILOT: TWR, Citation 1663, we have _____ due to a tire puncture, request hold on the runway and _____. No reported fire or smoke and we have shut down our engines as a precaution.
4. CONTROLLER: All aircraft, caution, _____ on RWY 16R reported medium on the first half of the runway and poor on the second half due to _____.
5. PILOT: TWR, OB Air 1663, request _____ for pushback due to no show passenger. We have to offload his _____.

PRONUNCIATION CD21
CDで上の文を聞き、後についてリピートしましょう。何回か繰り返します。

FLUENCY
上のそれぞれのATCの会話について、PROBLEM と RESPONSE を描写しながら話しましょう。

UNIT 4 CLEARED FOR TAKEOFF

SINGLE PICTURE

FLUENCY

絵の状況について、SPRCモデルを使って、3分程度話しましょう。まず、左下の文を使って話し始めます。次に、右下のマインドマップの言葉を使って話すようにします。それをS、P、R、Cとそれぞれ行います。

*Organize your ideas with the **SPRC** model!*

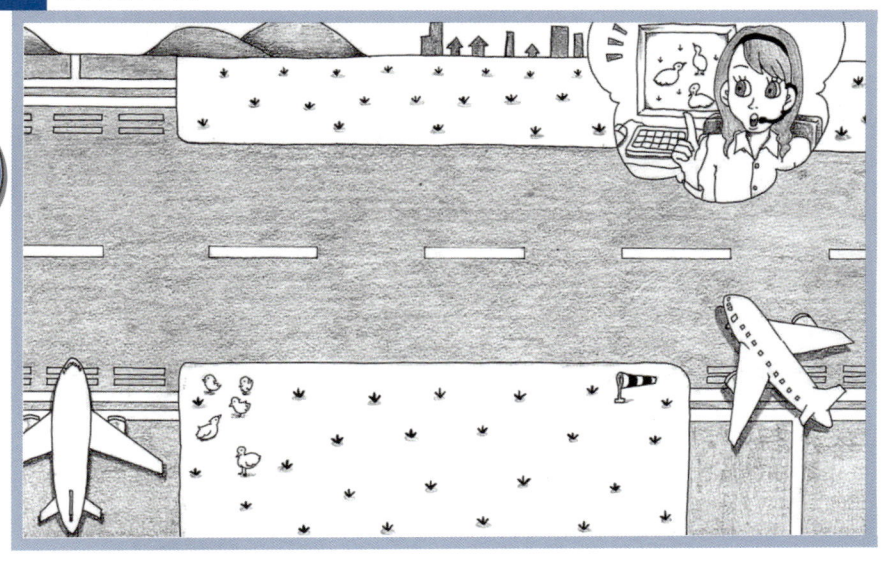

Speak in complete sentences!

S = SITUATION

In this picture I can see a runway, some taxiways and two airplanes…

- Runway area
  - how many airplanes?
  - doing what?
  - birds
- Controller
  - issue caution
  - all aircraft

P = PROBLEM

Birds are a hazard, especially at airports…

- Hazard
  - bird strikes
  - what kind of damage?

R = RESPONSE

Before takeoff there are several things that the pilots should do…

- Before takeoff
  - crew briefing
  - check & confirm
  - request bird sweep
- During takeoff
  - if bird strike…
  - reject takeoff
  - contact tower
- After takeoff
  - if bird strike…
  - emergency landing
  - tow assistance
  - fire equipment

C = CONCLUSION

Pilots must use extreme caution at airports where birds are common…

- Caution
  - be prepared
  - check & confirm
  - before & after takeoff

UNIT 4   CLEARED FOR TAKEOFF

USEFUL LANGUAGE

STRUCTURE

このエクササイズでは、助動詞の「must」と「has/have to」を使って、必要な行動を表わします。下の単語を並べ替えて文を作りましょう。1問目の例を参考にしてください。

NECESSARY ACTIONS

1. is / obstruction / the runway. / some kind of / There / on
   → *There is some kind of obstruction on the runway.*

2. due to / The pilots / takeoff / reject / have to / the obstruction.
   → _____

3. the tower / the incident. / The pilots / contact / and report / must
   → _____

4. two airplanes / the active / There are / runway. / on
   → _____

5. has to / the runway. / The small / clear / airplane
   → _____

6. takeoff / must / his position. / and hold / reject / The pilot
   → _____

7. seems / the fuselage. / leaking from / to be / There / fluid
   → _____

8. The pilots / check / the problem. / have to / to find / the aircraft systems
   → _____

9. the situation. / contact / and describe / The pilots / the controller / must
   → _____

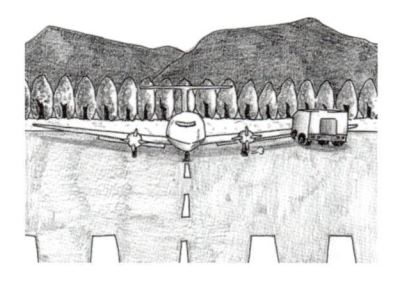

10. has / problem. / like / a landing gear / It looks / this aircraft
    → _____

11. must / air traffic / direct / go around. / to / The controller
    → _____

12. tow assistance / The pilot / the runway. / request / has to / to get off
    → _____

PRONUNCIATION

CD22 CDで上の文を聞き、後についてリピートしましょう。何回か繰り返します。(回答は81ページ)

FLUENCY

上の4つの状況と同じような経験があれば、描写しながら話しましょう。

UNIT 4   CLEARED FOR TAKEOFF

PICTURE SEQUENCE

Use "because" to give reasons!

INTERACTIONS

CD23 1～4の絵について以下の問題に答えましょう。まずは、CDでそれぞれの問題を聞きます。問題を聞いた後に一時停止ボタンを押して、それぞれの問題に対して回答します。完全文を使って、回答は声に出して言いましょう。

Picture 1

1. What was the situation at this airport?
2. What did the controller tell the pilots?

Picture 2

3. What was the problem?
4. Describe the options for the pilot.

Picture 3

5. What happened to the airplane?
6. What were the dangers in this situation?

Picture 4

7. How did the pilot respond?
8. What do you think happened next?

UNIT 4 CLEARED FOR TAKEOFF

VOCABULARY REVIEW

VOCABULARY

キーワードの復習です。ヒントを読んで、クロスワードに答えを書きましょう。(回答は 81 ページ)

ACROSS

1. the structure that supports an aircraft on the ground and allows it to taxi
4. a collision between a plane and a bird, usually during takeoff or landing
5. things that are used to put out a fire (eg: fire extinguisher, foam)
8. to change an event to a later time
10. using a ground vehicle to move an aircraft that has a problem

DOWN

2. a situation in which a pilot decides to abort takeoff
3. an event that happens during operations which affects aircraft safety
6. an aborted landing for an aircraft on final approach
7. a procedure to remove birds from an area of the airport
9. a controller or pilot must wait for a short time

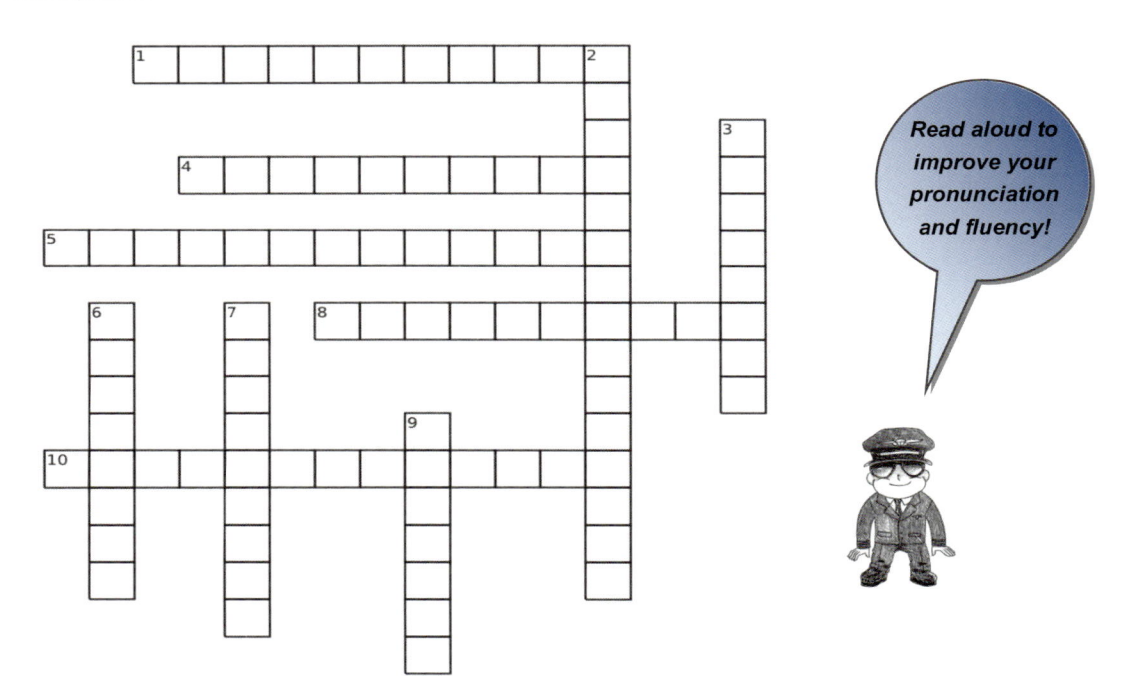

Read aloud to improve your pronunciation and fluency!

FLUENCY
SENTENCES

上のいくつかのキーワードを使って、3 つの文を作りましょう。その後文を何回か音読します。

1. _____

2. _____

3. _____

UNIT 4

CLEARED FOR TAKEOFF

SINGLE PICTURE REVIEW

VOCABULARY

以下のキーワードを使って、下線部を埋めましょう。(回答は 81 ページ)

KEY WORDS

a. bird strike
b. bird sweeps
c. fire equipment
d. go around
e. incidents
f. landing gear
g. reject takeoff
h. re-scheduled
i. stand by
j. tow assistance

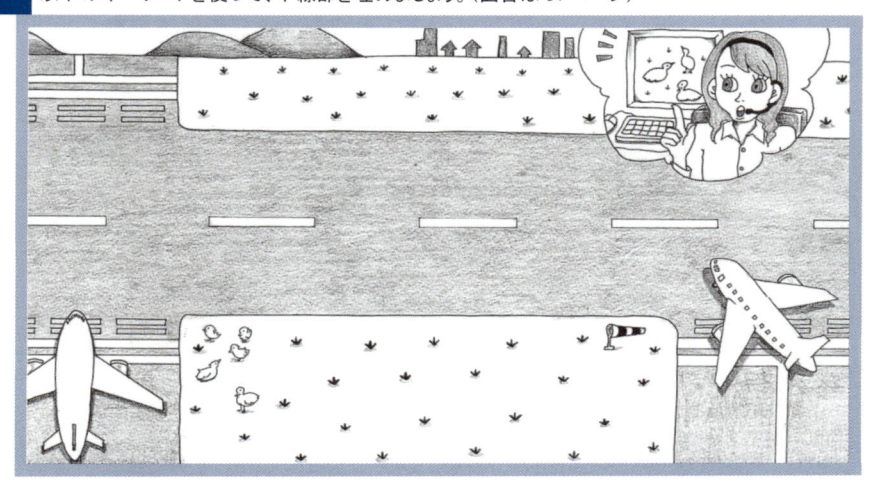

SITUATION

In this picture I can see a runway, some taxiways and two airplanes. One plane is entering the runway and the pilots are getting ready for takeoff. Another aircraft is holding on a taxiway. The pilot of this aircraft must __1._____ for clearance to cross the runway. I can also see several birds at the side of the runway. It looks like the controller is issuing a caution about birds to all aircraft.

PROBLEM

Birds are a hazard, especially at airports, and a lot of pilots have experienced bird strikes. A bird strike can damage the windshield, wing edges, __2._____ or engines of an aircraft. If a bird strike damages an engine, it can be very dangerous. Bird strikes are the cause of many __3._____ and accidents in aviation every year.

RESPONSE

Before takeoff there are several things that the pilots should do. First, the pilots should brief each other about a possible __4._____ during takeoff. Next, they should check carefully for birds in the area of the airport. Finally, if the captain sees many birds, he should request a bird sweep to clear the runway area. A bird strike during or after takeoff can be very dangerous. If there is a bird strike during takeoff, the pilots should __5._____ and contact the tower. If a bird strike happens after takeoff, the pilots have to make an emergency landing. The airplane may not be able to taxi, so the pilots may have to request __6._____. Occasionally a bird strike might result in an engine failure or fire. In this case, the captain has to request __7._____ and prepare for an evacuation. All arriving aircraft must __8._____ and stand by for further instructions. In addition, flights may be delayed, canceled or __9._____ due to airport closure.

CONCLUSION

Pilots must use extreme caution at airports where birds are common. __10._____ are often an effective way to scare birds away, but a bird strike can happen at any time and any place. Pilots must be prepared to take immediate action. They should check carefully for flocks of birds in the area, especially before and after takeoff.

FLUENCY

CD24 CDで上の文を聞き、音読しましょう。何回か音読を繰り返します。

UNIT 5 TAKEOFF & CLIMB

"CHECK AND CONFIRM"

AIR TURNBACK

VOCABULARY

それぞれのキーワードにあった意味を選びましょう。例を参考にしてください。(回答は 82 ページ)

KEY WORDS

1. air turnback — e
2. baggage compartment — ___
3. caution — ___
4. equipment failure — ___
5. flap — ___
6. injured passenger — ___
7. nose gear — ___
8. retract position — ___
9. stuck landing gear — ___
10. EICAS — ___

a. a situation where the undercarriage does not retract properly
b. a control surface which increases lift or drag
c. the forward part of the undercarriage
d. the setting where landing gear or flaps are pulled in or moved up
e. returning to the departure airport after takeoff due to a problem
f. a situation where airplane hardware is not working properly
g. Engine Indicating and Crew Alert System
h. an enclosed area in the cabin or cargo bay for storing bags
i. a traveler on an aircraft who becomes hurt
j. a warning to the pilot of a possible hazard or failure

PRONUNCIATION CD25
CDで上の文を聞き、後についてリピートしましょう。何回か繰り返します。

COMPREHENSION CD26
CDで5つのATCの会話を聞き、下線部を埋めましょう。(回答は 82 ページ)

LISTENING TO ATC

1. CONTROLLER: King Air 18C, _____, wake turbulence from _____ aircraft. Cleared for takeoff, RWY 18.
2. PILOT: TWR, King Air 18C, HDG 160, climbing through 1,200. We have a _____. Request _____ and low approach for gear check.
3. CONTROLLER: Citation 1663, hold over LIBRA, expect delay _____ or more. We have an aircraft with a _____ problem.
4. CONTROLLER: All aircraft, caution. _____ activity reported over the airport. Wind _____, gusting 40.
5. PILOT: Tokyo Ground, Cessna 1663, we are holding on Taxiway Bravo, unable to taxi due to _____ conditions and poor visibility. Request a tow back to the _____.

PRONUNCIATION CD27
CDで上の文を聞き、後についてリピートしましょう。何回か繰り返します。

FLUENCY
上のそれぞれのATCの会話について、PROBLEMとRESPONSEを描写しながら話しましょう。

UNIT 5 TAKEOFF & CLIMB

SINGLE PICTURE

FLUENCY

絵の状況について、SPRC モデルを使って、3 分程度話しましょう。まず、左下の文を使って話し始めます。次に、右下のマインドマップの言葉を使って話すようにします。それを S, P, R, C とそれぞれ行います。

*Organize your ideas with the **SPRC** model!*

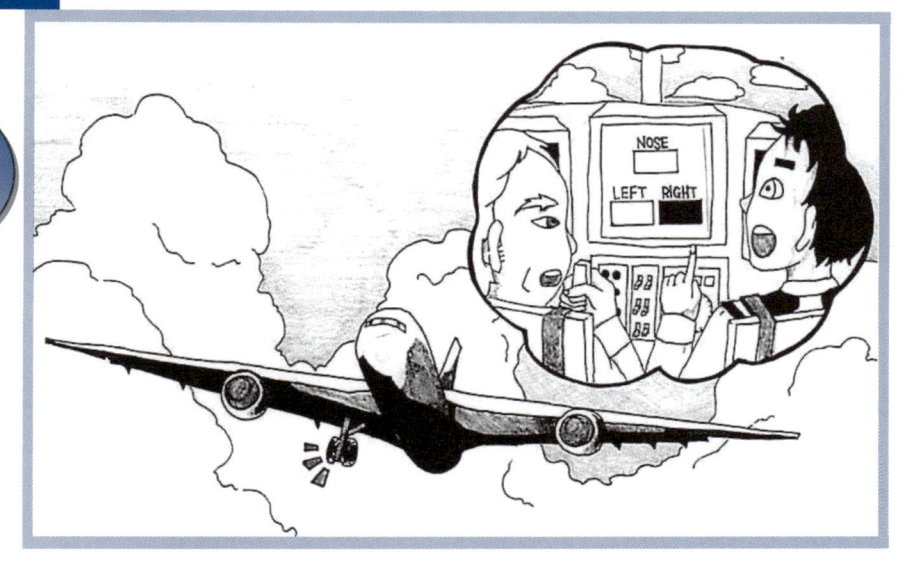

Use present tenses for the single picture!

S = SITUATION

In this picture I can see a twin-jet airliner…

Aircraft
- what type?
- climb
- sky condition

Cockpit
- captain
- first officer
- EICAS

P = PROBLEM

There seems to be some kind of equipment failure…

Equipment failure
- what equipment?
- what kind of danger?

R = RESPONSE

There are several things that the pilots should do…

Pilots
- EICAS & checklists
- contact ATC
- what kind of request?

Cabin crew
- secure cabin
- inform passengers
- calm passengers

C = CONCLUSION

In this situation, the most important thing is the safety of the passengers…

Safety
- risk of fire or injury
- emergency equipment
- tow assistance

UNIT 5

TAKEOFF & CLIMB

USEFUL LANGUAGE

STRUCTURE

このエクササイズでは、現在進行形の「is/are VERB+ing」を使って、今起きている事を表わします。下の単語を並べ替えて文を作りましょう。1問目の例を参考にしてください。

EVENTS HAPPENING NOW

1. is climbing / the airport. / from / The airplane / away
   → _The airplane is climbing away from the airport._
2. trying / is / the flaps. / to retract / The captain
   → _____
3. and fix / must / The captain / the airport / return to / the problem.
   → _____
4. is falling. / is / and snow / The weather / very cold 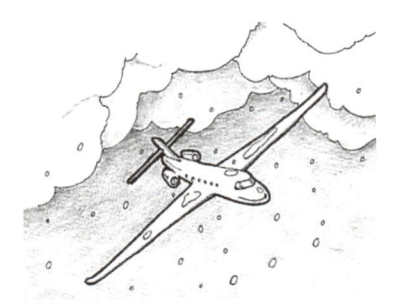
   → _____
5. the wings / is / and fuselage. / building up / Ice / on
   → _____
6. The captain / out of / fly / conditions. / must / the icing
   → _____
7. an equipment / experiencing / The aircraft / failure. / is 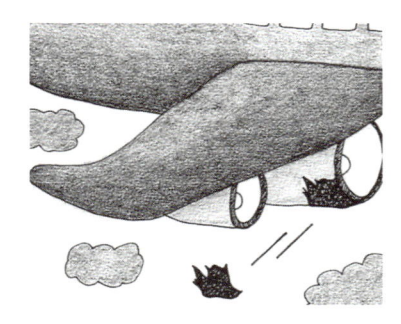
   → _____
8. is / from / the engines. / falling / one of / A panel
   → _____
9. an air turnback. / contact / The captain / and request / should / ATC
   → _____
10. his chest / looks sick. / is / and / One passenger / holding 
    → _____
11. A cabin / is / the captain. / notifying / attendant
    → _____
12. must / medical attention. / passenger / immediate / The sick / receive
    → _____

PRONUNCIATION

CD28 CDで上の文を聞き、後についてリピートしましょう。何回か繰り返します。(回答は82ページ)

FLUENCY

上の4つの状況と同じような経験があれば、描写しながら話しましょう。

UNIT 5   TAKEOFF & CLIMB

PICTURE SEQUENCE

Use past tenses for the picture sequence!

INTERACTIONS

CD29 1～4の絵について以下の問題に答えましょう。まずは、CDでそれぞれの問題を聞きます。問題を聞いた後に一時停止ボタンを押して、それぞれの問題に対して回答します。完全文を使って、回答は声に出して言いましょう。

Picture 1
1. What was the situation?
2. What were the pilots doing?

Picture 2
3. What was the problem after takeoff?
4. Describe the options for the pilots.

Picture 3
5. How did the crew prepare for landing?
6. How did the passengers respond to the problem?

Picture 4
7. What were the pilots thinking before landing?
8. What do you think happened after landing?

UNIT 5 TAKEOFF & CLIMB

VOCABULARY REVIEW

VOCABULARY

キーワードの復習です。ヒントを読んで、クロスワードに答えを書きましょう。(回答は 82 ページ)

ACROSS

3. a situation where airplane hardware is not working properly
4. Engine Indicating and Crew Alert System
5. the forward part of the undercarriage
8. a traveler on an aircraft who becomes hurt
10. a situation where the undercarriage does not retract properly

DOWN

1. the setting where landing gear or flaps are pulled in or moved up
2. an enclosed area in the cabin or cargo bay for storing bags
6. returning to the departure airport after takeoff due to a problem
7. a warning to the pilot of a possible hazard or failure
9. a control surface which increases lift or drag

To learn new words, it is important to use them many times!

FLUENCY

上のいくつかのキーワードを使って、3つの文を作りましょう。その後文を何回か音読します。

SENTENCES

1. _____

2. _____

3. _____

UNIT 5

TAKEOFF & CLIMB

SINGLE PICTURE REVIEW

VOCABULARY

以下のキーワードを使って、下線部を埋めましょう。(回答は 82 ページ)

KEY WORDS

a. air turnback
b. baggage compartments
c. caution
d. equipment failure
e. flaps
f. injured passengers
g. nose gear
h. retract position
i. stuck landing gear
j. EICAS

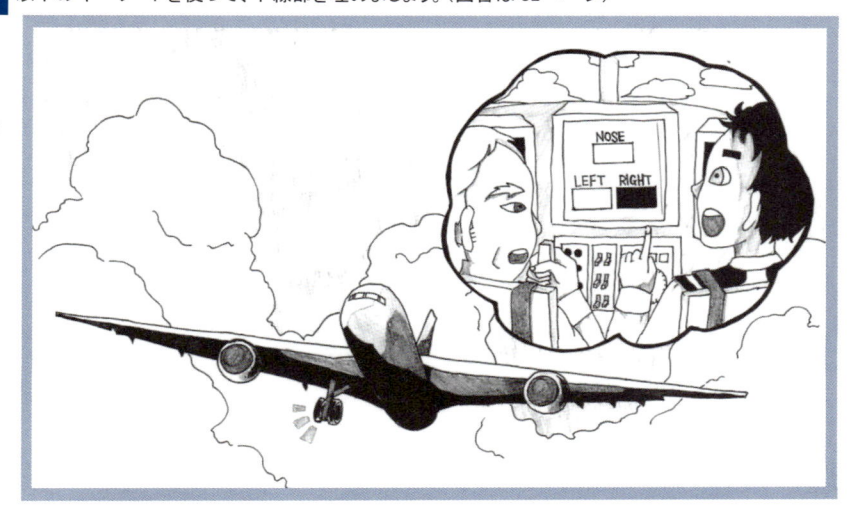

SITUATION

In this picture I can see a twin-jet airliner. The aircraft is climbing after takeoff. I think the weather looks okay. The sky condition is scattered because I can see blue sky and a few clouds behind the plane. In the cockpit there are two pilots. The captain is in the left seat and he is flying the aircraft. The first officer is in the right seat. He is looking at a __1._____ on the EICAS screen.

PROBLEM

There seems to be some kind of equipment failure. I think that there is a problem with the landing gear. After takeoff the pilots set the __2._____ and landing gear to the __3._____. Now the EICAS screen indicates that the __4._____ and left landing gear is retracted. However, it also indicates that the right landing gear is still extended. This is a problem because the captain cannot continue the flight with __5._____.

RESPONSE

There are several things that the pilots should do. First, they should try to fix the problem using the __6._____ system and checklists. If the problem cannot be fixed, the pilots must contact the air traffic controller and request an __7._____. They may also request a low approach over the runway. Then the controller can look closely at the landing gear to see if it is safe for landing. In addition, the captain has to inform the cabin crew of the situation. The cabin crew must secure the cabin by checking the __8._____ and checking that passengers' seatbelts are fastened. The crew must also inform the passengers that the airplane is returning to the departure airport due to an __9._____. Finally, the cabin crew should try to keep the passengers calm and relaxed.

CONCLUSION

In this situation, the most important thing is the safety of the passengers and crew. It may be difficult to make a smooth landing, and there is a risk of fire or __10._____. Therefore, the aircraft may require emergency equipment or tow assistance after landing. Later the passengers will be rescheduled to other flights, and the landing gear must be inspected by the maintenance crew.

FLUENCY

CD30 CDで上の文を聞き、音読しましょう。何回か音読を繰り返します。

UNIT 6 — CLIMB

"LEVEL OFF AT YOUR DISCRETION"
EQUIPMENT FAILURE & ROUGH AIR

VOCABULARY
それぞれのキーワードにあった意味を選びましょう。例を参考にしてください。(回答は 83 ページ)

KEY WORDS

1. ash cloud — g
2. landing area — ___
3. level off — ___
4. malfunction — ___
5. non-normal situation — ___
6. terrain — ___
7. visual check — ___
8. volcanic eruption — ___
9. windshield — ___
10. PIREP — ___

a. the physical features of an area of land
b. the window at the front of the cockpit
c. looking and confirming something by sight
d. a place where an aircraft can land
e. a pilot report about a flight hazard or dangerous encounter
f. to fail to work properly
g. a large amount of small particles thrown into the air by a volcano
h. when something unusual happens
i. to stop climbing or descending and maintain an assigned altitude
j. an explosion of dust, rocks and flames from a volcano

PRONUNCIATION
CD31 CDで上の文を聞き、後についてリピートしましょう。何回か繰り返します。

COMPREHENSION
CD32 CDで 5 つのATCの会話を聞き、下線部を埋めましょう。(回答は 83 ページ)

LISTENING TO ATC

1. CONTROLLER: Citation 1663, we've received a _____ from a 737 reporting volcanic _____ encounter at FL170, 30 miles southeast of Sakurajima.
2. PILOT: TWR, Citation 1663, FL 250, request _____ due to a _____ who needs immediate medical assistance.
3. PILOT: TWR, Citation 1663, now leveled off at FL200. We encountered _____ between _____. No known damage to aircraft or injuries to passengers.
4. PILOT: TWR, Citation 1663, we have a _____. Our _____ will not retract. Request hold over DELTA so that we can troubleshoot the problem.
5. CONTROLLER: All aircraft, _____ reported by a Citation during climb between 8,000 and 10,000 approximately _____ of the airport.

PRONUNCIATION
CD33 CDで上の文を聞き、後についてリピートしましょう。何回か繰り返します。

FLUENCY
上のそれぞれのATCの会話について、PROBLEMとRESPONSEを描写しながら話しましょう。

UNIT 6 — CLIMB

SINGLE PICTURE

FLUENCY

絵の状況について、SPRCモデルを使って、3分程度話しましょう。まず、左下の文を使って話し始めます。次に、右下のマインドマップの言葉を使って話すようにします。それをS、P、R、Cとそれぞれ行います。

*Organize your ideas with the **SPRC** model!*

S = SITUATION

In this picture I can see a light aircraft flying…

- Aircraft
  - what type?
  - purpose of flight
- Terrain
  - mountains
  - active volcano
  - ash cloud
- Controller
  - caution
  - volcano

P = PROBLEM

An ash cloud encounter is a very dangerous non-normal situation…

- Ash cloud encounter
  - non-normal situation
  - what kind of damage?
  - difficult to detect

Make sentences with the words in the mindmap!

R = RESPONSE

There are several ways in which a pilot can respond to this problem…

- Pilot responses
  - avoid area
  - PIREP to ATC
  - emergency landing

C = CONCLUSION

Pilots must use extreme caution when flying in areas with active volcanoes…

- Flight planning
  - weather reports
  - PIREPs
- During flight
  - visual check
  - be aware of danger

UNIT 6 — CLIMB

USEFUL LANGUAGE

STRUCTURE

このエクササイズでは、過去進行形と単純過去形を使って、過去の予期せぬ出来事を表わします。下の単語を並べ替えて文を作りましょう。1問目の例を参考にしてください。

INTERRUPTED ACTIONS IN THE PAST

1. it encountered / clouds / turbulence. / was entering / when / The plane
   → *The plane was entering clouds when it encountered turbulence.*

2. was / The pilot / a passenger / levelling off / was injured. / when
   → _____

3. the CAs / about / the captain / One of / the injured passenger. / notified
   → _____

4. overheated. / was / the engine / A Cessna / when / changing heading
   → _____

5. was / the engine stopped. / looking for / when / The pilot / a landing area
   → _____

6. and landed / The pilot / on / an emergency / a golf course. / declared
   → _____

7. entering clouds / icing conditions. / The plane / it encountered / was / when
   → _____

8. Ice / made / on the wings / a visual check. / when the crew / was building up
   → _____

9. to / The captain / activate / anti-icing systems. / had / the aircraft's
   → _____

10. climbing / opened. / The plane / when / was / a baggage compartment
    → _____

11. to his wife / he was hit / was talking / by a bag. / when / A passenger
    → _____

12. Cabin crew / give / to / medical assistance / the passenger. / had to
    → _____

PRONUNCIATION

CD34 CDで上の文を聞き、後についてリピートしましょう。何回か繰り返します。(回答は83ページ)

FLUENCY

上の4つの状況と同じような経験があれば、描写しながら話しましょう。

UNIT 6 CLIMB

PICTURE SEQUENCE

Try to make accurate word choices!

INTERACTIONS

CD35 1～4の絵について以下の問題に答えましょう。まずは、CDでそれぞれの問題を聞きます。問題を聞いた後に一時停止ボタンを押して、それぞれの問題に対して回答します。完全文を使って、回答は声に出して言いましょう。

Picture 1
1. What was the situation?
2. What was the controller doing?

Picture 2
3. What was the problem?
4. How did the pilot respond?

Picture 3
5. Describe the passengers.
6. What did the pilot say to them?

Picture 4
7. Describe the landing.
8. What do you think happened next?

UNIT 6   CLIMB

VOCABULARY REVIEW

VOCABULARY

キーワードの復習です。ヒントを読んで、クロスワードに答えを書きましょう。(回答は 83 ページ)

ACROSS

5. an explosion of dust, rocks and flames from a volcano
6. to fail to work properly
7. a large amount of small particles thrown into the air by a volcano
9. the physical features of an area of land
10. a place where an aircraft can land

DOWN

1. when something unusual happens
2. a pilot report about a flight hazard or dangerous encounter
3. the window at the front of the cockpit
4. looking and confirming something by sight
8. to stop climbing or descending and maintain an assigned altitude

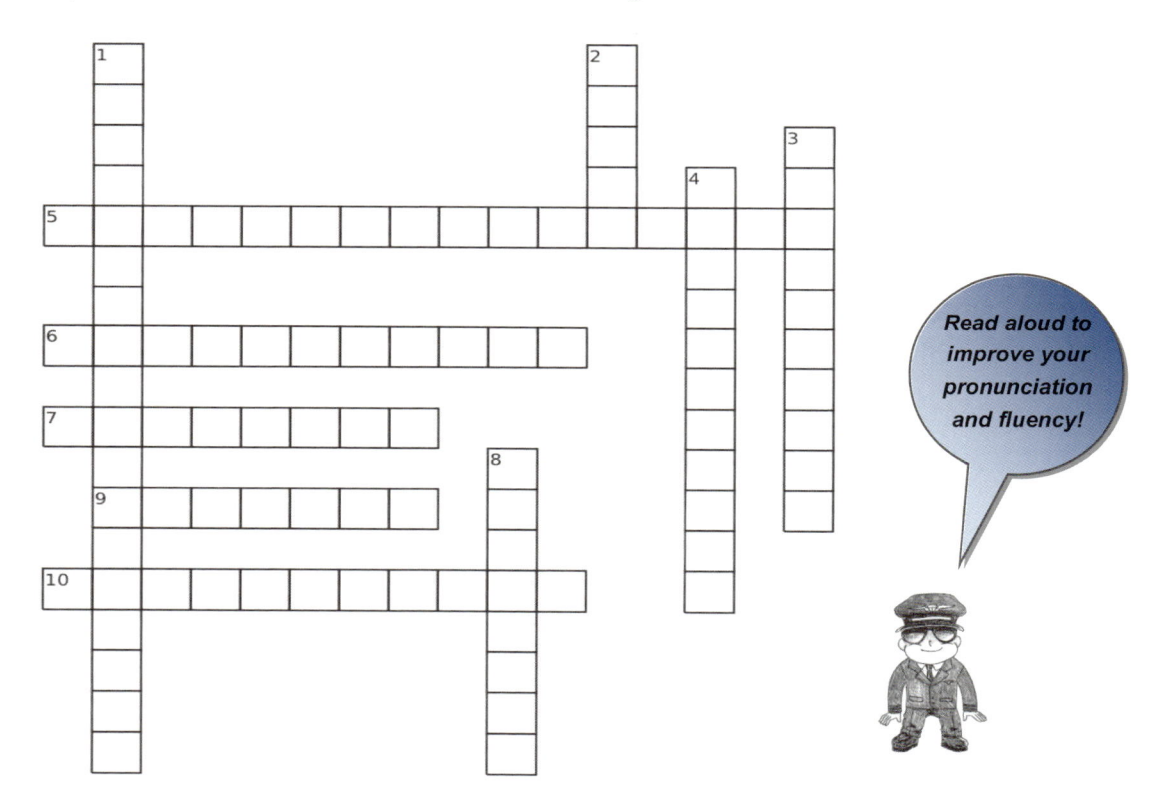

Read aloud to improve your pronunciation and fluency!

FLUENCY

SENTENCES

上のいくつかのキーワードを使って、3つの文を作りましょう。その後文を何回か音読します。

1. _____
2. _____
3. _____

UNIT 6 — CLIMB

SINGLE PICTURE REVIEW

VOCABULARY

以下のキーワードを使って、下線部を埋めましょう。(回答は 83 ページ)

KEY WORDS

a. ash cloud
b. landing area
c. leveling off
d. malfunction
e. non-normal situation
f. terrain
g. visual check
h. volcanic eruptions
i. windshield
j. PIREP

SITUATION

In this picture I can see a light aircraft flying in mountain __1._____. It looks like the pilot is completing his climb and is __2._____. I can see the pilot and some passengers. The visibility is good and I think this might be a sightseeing tour. I can also see a large ash cloud coming from a volcano. Maybe the pilot is trying to fly close to look at the active volcano. A controller is issuing a caution. I think that she is warning all aircraft about the active volcano.

PROBLEM

An ash cloud encounter is a very dangerous __3._____ for pilots. Ash clouds often form after __4._____. The ash is like fine dust and can cause a lot of damage to aircraft. It can scratch the __5._____ and make it difficult for the pilot to see. In the case of jet aircraft, ash may enter the jet engines and cause them to __6._____. Another problem is detecting an ash cloud. On cloudy days it is difficult to see ash clouds. At night, it may be impossible to detect an ash cloud encounter until it is too late.

RESPONSE

There are several ways in which a pilot can respond to this problem. First, pilots should always avoid the area around active volcanoes. This pilot is flying too close to the volcano. He should avoid this area. Second, if a pilot sees a volcanic __7._____, he should make a __8._____ to the ATC and stay clear of the area. Finally, if an aircraft encounters an ash cloud, the pilot must quickly find a way out. In extreme cases, the pilot may have to find a safe __9._____ and make an emergency landing.

CONCLUSION

Pilots must use extreme caution when flying in areas with active volcanoes. Flight planning should include a check of the latest weather reports and PIREPs of ash cloud activity. During flight, pilots should keep a __10._____ of areas with possible volcanic ash activity. At all times pilots must be aware of the dangers of flying into ash clouds.

FLUENCY

CD36 CDで上の文を聞き、音読しましょう。何回か音読を繰り返します。

UNIT 7   CRUISE

"Is There A Doctor On Board?"
PASSENGER INJURIES & PROBLEMS

VOCABULARY

それぞれのキーワードにあった意味を選びましょう。例を参考にしてください。（回答は 84 ページ）

KEY WORDS

1. chief purser — i
2. destination — ___
3. diversion — ___
4. divert — ___
5. drunk — ___
6. evacuation — ___
7. medical assistance — ___
8. police assistance — ___
9. restrain — ___
10. unruly behavior — ___

a. to change from the planned course to another course
b. when a passenger disrupts cabin crew and other passengers
c. to keep a person under control, for example by using plastic loops
d. quickly removing all passengers and crew from an aircraft
e. the place designated as the end of a flight
f. police officers helping the crew deal with unruly passengers
g. changing the planned course, for example due to weather
h. unable to speak or act normally due to drinking too much alcohol
i. the cabin attendant in charge of the cabin
j. doctors or medics helping a sick or injured passenger

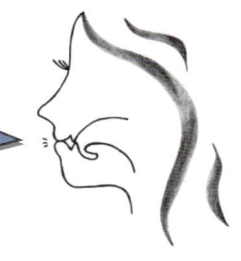

PRONUNCIATION   CD37
CDで上の文を聞き、後についてリピートしましょう。何回か繰り返します。

PRONUNCIATION   CD38
CDで以下の表現を聞き、後についてリピートしましょう。

"B" IN VARIOUS POSITIONS

1. push**b**ack
2. runway num**b**er
3. micro**b**urst
4. air turn**b**ack
5. noise a**b**atement
6. departure lo**bb**y
7. am**b**ulance
8. ca**b**in attendant
9. wake tur**b**ulence
10. porta**b**le **b**oarding stairs

To make the "B" sound: put your lips together, then open them quickly and use your voice!

PRONUNCIATION   CD39
CDで以下の表現を聞き、後についてリピートしましょう。

"V" IN VARIOUS POSITIONS

1. acti**v**e runway
2. le**v**el off
3. see and a**v**oid
4. re**v**erse thrust
5. ad**v**erse weather
6. flight le**v**el 250
7. ci**v**il a**v**iation
8. runway o**v**errun
9. **V**FR na**v**igation
10. **v**olcanic ash acti**v**ity

To make the "V" sound: put your top front teeth and bottom lip together, then open your mouth and use your voice!

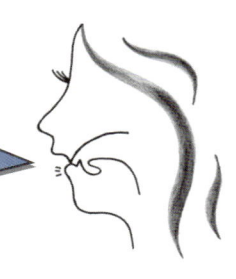

UNIT 7   CRUISE

SINGLE PICTURE

FLUENCY   絵の状況について、SPRCモデルを使って、3分程度話しましょう。まず、左下の文を使って話し始めます。次に、右下のマインドマップの言葉を使って話すようにします。それをS、P、R、Cとそれぞれ行います。

*Organize your ideas with the **SPRC** model!*

Use present tenses for the single picture!

S = SITUATION

This picture shows the cabin of a large airliner…

Airliner cabin
- cruise
- cabin crew
- passengers

P = PROBLEM

One male passenger is causing a problem…

Alcohol
- drunk passenger
- beer bottle
- what kind of danger?

Smoking
- cigarette
- what kind of danger?

Unruly behavior
- upset other passengers

R = RESPONSE

The cabin crew and pilots must work together to control this situation…

Cabin crew
- notify pilots
- warn man
- restrain man
- calm passengers

Pilots
- divert or continue?
- contact company & ATC
- what kind of request?

C = CONCLUSION

It is important to control an unruly passenger as soon as possible…

Control situation
- restrain unruly passenger
- possible diversion
- crew coordination & training

UNIT 7

UNIT 7   CRUISE

USEFUL LANGUAGE

STRUCTURE

このエクササイズでは、助動詞の「should」を使って、問題解決のアドバイスを表わします。下の単語を並べ替えて文を作りましょう。1問目の例を参考にしてください。

SOLVING PROBLEMS

1. should tell / it hurts. / exactly where / The boy / the CA
   → *The boy should tell the CA exactly where it hurts.*

2. inform / the sick passenger. / about / should / the pilots / The chief purser
   → _____

3. medical assistance. / make / The captain / a doctor call / should / for
   → _____

4. to / explain / The husband / the situation / the CA. / should
   → _____

5. the pregnant passenger / should move / a comfortable seat. / The CA / to
   → _____

6. contact / for / The captain / advice. / should / the company
   → _____

7. not play / his computer. / should / loud music / The passenger / on
   → _____

8. the boy / The CA / to turn down / politely ask / should / the volume.
   → _____

9. put away / takeoff and landing. / The boy / his computer / should / before
   → _____

10. should / hospital. / The medics / the sick passenger / quickly take / to
    → _____

11. The captain / stay calm. / tell / to / the other passengers / should
    → _____

12. not / The other passengers / their seats / should / yet. / leave
    → _____

PRONUNCIATION

CD40 CDで上の文を聞き、後についてリピートしましょう。何回か繰り返します。(回答は84ページ)

FLUENCY

上の4つの状況と同じような経験があれば、描写しながら話しましょう。

UNIT 7 — CRUISE

PICTURE SEQUENCE

Use past tenses for the picture sequence!

INTERACTIONS

CD41 1～4の絵について以下の問題に答えましょう。まずは、CDでそれぞれの問題を聞きます。問題を聞いた後に一時停止ボタンを押して、それぞれの問題に対して回答します。完全文を使って、回答は声に出して言いましょう。

Picture 1
1. Describe the situation.
2. What was the pilot doing?

Picture 2
3. What was the cabin attendant doing?
4. What problem happened?

Picture 3
5. How did the CA respond to the problem?
6. Who helped the CA?

Picture 4
7. What options did the pilots have?
8. What do you think happened next?

UNIT 7　　　　　　　　　　　　　　　　　　　　　　　　　　　　　　　CRUISE

VOCABULARY REVIEW

VOCABULARY

キーワードの復習です。ヒントを読んで、クロスワードに答えを書きましょう。（回答は 84 ページ）

ACROSS

3. the place designated as the end of a flight
7. police officers helping the crew deal with unruly passengers
8. to keep a person under control, for example by using plastic loops
9. unable to speak or act normally due to drinking too much alcohol
10. changing the planned course, for example due to weather

DOWN

1. doctors or medics helping a sick or injured passenger
2. when a passenger disrupts cabin crew and other passengers
4. the cabin attendant in charge of the cabin
5. quickly removing all passengers and crew from an aircraft
6. to change from the planned course to another course

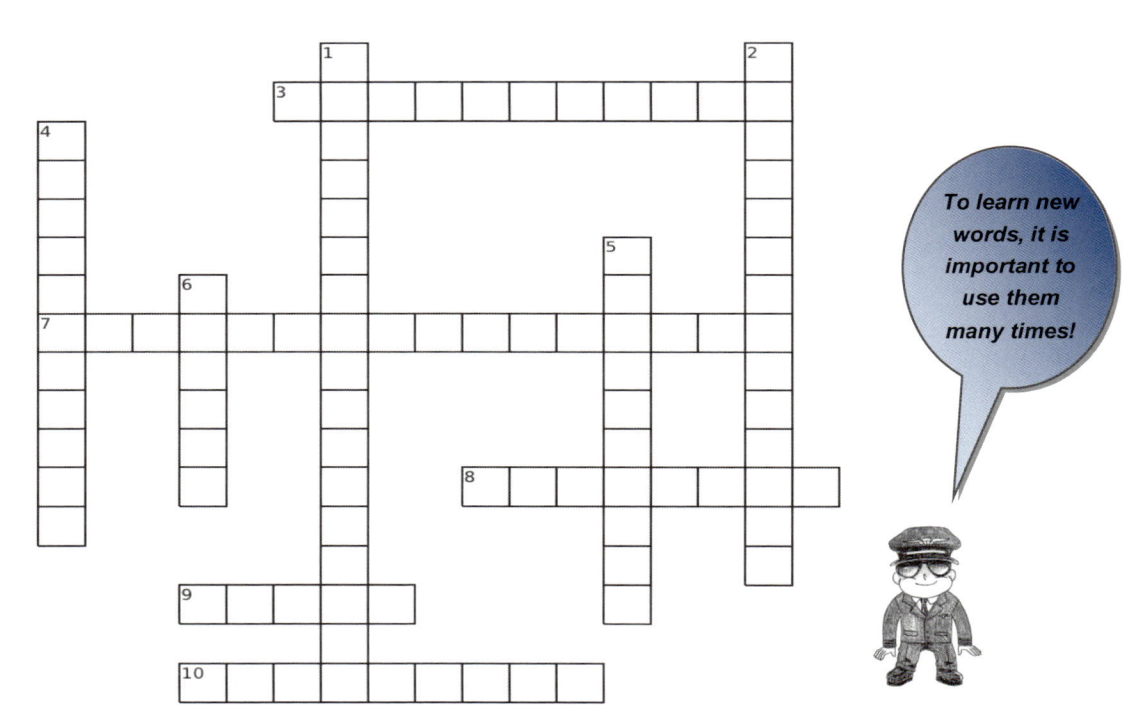

To learn new words, it is important to use them many times!

FLUENCY

上のいくつかのキーワードを使って、3 つの文を作りましょう。その後文を何回か音読します。

SENTENCES

1. _____

2. _____

3. _____

UNIT 7   CRUISE

SINGLE PICTURE REVIEW

VOCABULARY

以下のキーワードを使って、下線部を埋めましょう。(回答は 84 ページ)

KEY WORDS

a. chief purser
b. destination
c. diversion
d. divert
e. drunk
f. evacuation
g. medical assistance
h. police assistance
i. restrain
j. unruly behavior

SITUATION

This picture shows the cabin of a large airliner. There is a __1._____ and another female cabin attendant. I think this is cruise flight because the cabin crew are walking around and some passengers are not wearing seatbelts. I can see several passengers in the cabin and they all look upset. A cabin attendant is comforting a child who is crying.

PROBLEM

One male passenger is causing a problem. I think he has had too much alcohol. He seems to be __2._____ and he is waving a beer bottle in his left hand. If he injures other passengers with the bottle, they may need __3._____. He is also smoking a cigarette. Smoking during flight is prohibited because it is dangerous. If there is a fire in the cabin the captain will have to make an emergency landing and __4._____. In addition, this passenger has his right foot on the armrest of the seat, and his arm is around another passenger. His __5._____ is upsetting the other passengers.

RESPONSE

The cabin crew and pilots must work together to control this situation. First, the cabin crew should notify the pilots about the problem. Then the chief purser should warn the man to stop smoking cigarettes and drinking alcohol. If he continues to behave badly, the cabin crew must __6._____ him. The unruly passenger looks big and strong so the chief purser may ask other male passengers for help. They should move him away from other passengers so that he will not cause injury. Then the cabin attendants have to calm down the other passengers. Also, the captain must decide either to __7._____ to the nearest airport or continue to the original __8._____. He should contact the company and ATC to inform them about the situation. Finally, the captain may request __9._____ after landing.

CONCLUSION

It is important to control an unruly passenger as soon as possible. In some cases the cabin crew may have to restrain a passenger. If that happens, the captain should declare an emergency and make a __10._____ to a nearby airport. Good crew coordination and training is important to handle this type of dangerous situation.

FLUENCY   CD42 CDで上の文を聞き、音読しましょう。何回か音読を繰り返します。

UNIT 8 — EMERGENCY

"MAYDAY, MAYDAY, MAYDAY!"
SMOKE IN THE CABIN

VOCABULARY

それぞれのキーワードにあった意味を選びましょう。例を参考にしてください。（回答は 85 ページ）

KEY WORDS

1. alternate airport — h
2. controlled descent — ___
3. cruise flight — ___
4. emergency procedure — ___
5. flight plan — ___
6. oxygen mask — ___
7. priority landing — ___
8. radar vectors — ___
9. rapid decompression — ___
10. warning — ___

a. the flight phase after climb and before descent
b. a situation where an aircraft moves to the top of the traffic pattern
c. a sudden loss of cabin pressure
d. a device worn over the face that supplies oxygen
e. an alarm or light that indicates a problem or danger
f. a steady decrease in altitude from cruise flight
g. a plan of actions to be done in an emergency situation
h. a place where a plane may land if the intended airport is unavailable
i. directional instructions given by a controller to a pilot
j. a detailed schedule of the route for an aircraft

PRONUNCIATION

CD43 CDで上の文を聞き、後についてリピートしましょう。何回か繰り返します。

PRONUNCIATION

CD44 CDで以下の表現を聞き、後についてリピートしましょう。

"S" IN VARIOUS POSITIONS

1. callsign
2. supersonic
3. unsafe
4. responsibility
5. hydraulic system
6. sick passenger
7. Cessna 172
8. distress signal
9. search and rescue
10. overseas operations

To make the "S" sound: put the tip of your tongue behind the bottom front teeth!

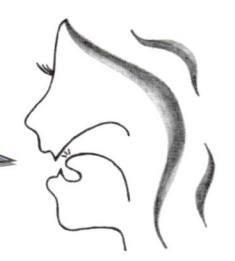

PRONUNCIATION

CD45 CDで以下の表現を聞き、後についてリピートしましょう。

"TH" IN VARIOUS POSITIONS

1. northeast
2. health
3. southwest
4. flight path
5. maximum thrust
6. earthquake
7. bandwidth
8. southbound
9. thunderstorm
10. authorized approach

To make the "TH" sound: put the tip of your tongue between the top and bottom front teeth!

UNIT 8 — EMERGENCY

SINGLE PICTURE

FLUENCY

絵の状況について、SPRC モデルを使って、3 分程度話しましょう。まず、左下の文を使って話し始めます。次に、右下のマインドマップの言葉を使って話すようにします。それを S、P、R、C とそれぞれ行います。

*Organize your ideas with the **SPRC** model!*

Speak in complete sentences!

S = SITUATION

This picture shows the inside of an airliner cabin…

Airliner cabin
- cruise
- meal service
- passengers

P = PROBLEM

There seems to be a serious problem in the cabin…

Cabin
- oxygen masks
- overhead baggage compartment

Rapid decompression
- air pressure drop
- mist
- cold & windy

R = RESPONSE

It is important to respond quickly to this kind of emergency…

Emergency procedure
- warning to pilots
- controlled descent
- contact ATC

Cabin crew
- secure cabin
- help passengers

C = CONCLUSION

In emergency situations the crew must work quickly and calmly…

Emergency situations
- cabin crew actions
- pilot actions

UNIT 8

EMERGENCY

USEFUL LANGUAGE

STRUCTURE

このエクササイズでは、「will」と「going to」を使って、未来の出来事を表わします。下の単語を並べ替えて文を作りましょう。1問目の例を参考にしてください。

ACTIONS IN THE FUTURE

1. is / a fire extinguisher. / to get / The cabin attendant / going
    → _The cabin attendant is going to get a fire extinguisher._
2. the fire extinguisher / put out / will use / to / She / the fire.
    → _____
3. the emergency. / inform / The chief purser / the captain / will / about
    → _____
4. is going / the passenger / The CA / to put away / to ask / the bottle. 
    → _____
5. are / the drunk man. / going / about / Other passengers / to complain
    → _____
6. the company / The captain / the passenger. / will ask / about / for advice
    → _____
7. are / of the passengers / going / oxygen masks. / to put on / All 
    → _____
8. ATC / an emergency landing. / contact / will / and request / The pilot
    → _____
9. going / evacuation. / The CAs / an emergency / to prepare for / are
    → _____
10. the airplane. / the passengers / going / All of / are / to evacuate 
    → _____
11. The CAs / assist / and children / will / to leave. / old passengers
    → _____
12. help / the plane. / will / Emergency staff / outside / injured passengers
    → _____

PRONUNCIATION

CD46 CDで上の文を聞き、後についてリピートしましょう。何回か繰り返します。(回答は85ページ)

FLUENCY

上の4つの状況と同じような経験があれば、描写しながら話しましょう。

UNIT 8 EMERGENCY

PICTURE SEQUENCE

Avoid using fillers and reduce your pauses!

INTERACTIONS

CD47 1〜4の絵について以下の問題に答えましょう。まずは、CDでそれぞれの問題を聞きます。問題を聞いた後に一時停止ボタンを押して、それぞれの問題に対して回答します。完全文を使って、回答は声に出して言いましょう。

Picture 1
1. Describe this place.
2. What were the cabin attendants doing?

Picture 2
3. What was happening in the cabin?
4. What problem did some passengers notice?

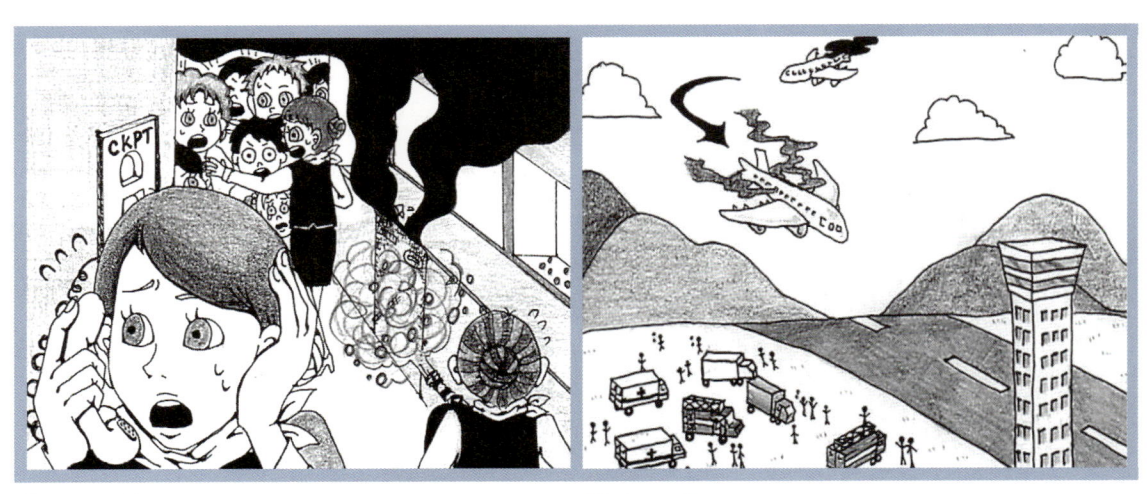

Picture 3
5. What happened in the galley?
6. How did the CAs respond to the problem?

Picture 4
7. How did the pilots respond to the problem?
8. What do you think happened next?

UNIT 8   EMERGENCY

VOCABULARY REVIEW

VOCABULARY

キーワードの復習です。ヒントを読んで、クロスワードに答えを書きましょう。(回答は 85 ページ)

ACROSS

2. the flight phase after climb and before descent
4. a place where a plane may land if the intended airport is unavailable
5. a device worn over the face that supplies oxygen
7. directional instructions given by a controller to a pilot
8. a sudden loss of cabin pressure
9. a situation where an aircraft moves to the top of the traffic pattern

DOWN

1. a plan of actions to be done in an emergency situation
2. a steady decrease in altitude from cruise flight
3. a detailed schedule of the route for an aircraft
6. an alarm or light that indicates a problem or danger

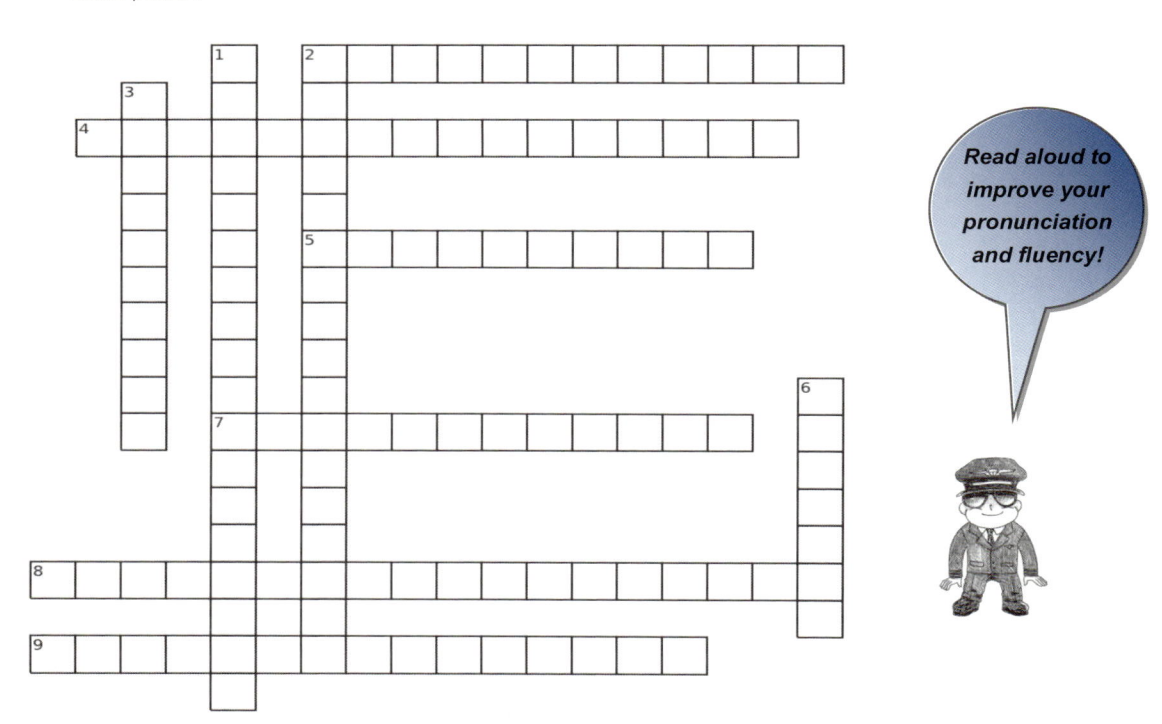

Read aloud to improve your pronunciation and fluency!

FLUENCY

上のいくつかのキーワードを使って、3つの文を作りましょう。その後文を何回か音読します。

SENTENCES

1. _____

2. _____

3. _____

UNIT 8   EMERGENCY

SINGLE PICTURE REVIEW

VOCABULARY

以下のキーワードを使って、下線部を埋めましょう。(回答は 85 ページ)

KEY WORDS

a. alternate airport
b. controlled descent
c. cruise flight
d. emergency procedure
e. flight plan
f. oxygen masks
g. priority landing
h. radar vectors
i. rapid decompression
j. warning

SITUATION

This picture shows the inside of an airliner cabin. I can see one cabin attendant and five passengers. It looks like the aircraft is in __1._____ because the cabin attendant is doing meal service. The tray tables are down and the passengers are now eating a meal.

PROBLEM

There seems to be a serious problem in the cabin. I can see several __2._____ and the passengers look very worried. The cabin attendant is looking at the overhead baggage compartments. It looks like one of the compartments is damaged. I think that this airplane may be experiencing a __3._____. When that happens, the air pressure drops suddenly and mist forms inside the plane. It becomes very cold and windy, and it is difficult for passengers and crew to breathe. In addition, the pilots may have difficulty controlling the aircraft.

RESPONSE

It is important to respond quickly to this kind of emergency. The flight crew should follow the __4._____ for rapid decompression. First the chief purser should contact the pilots and give a __5._____ about the rapid decompression. Then the captain will change his __6._____ and make a rapid controlled descent to a lower altitude. The pilots must put on oxygen masks and contact ATC about the emergency. They have to request a __7._____ at an alternate airport so that they can land as soon as possible. The ATC controller will acknowledge the emergency and give __8._____ to the nearest airport. Meanwhile the cabin crew should secure loose items, and they must check that the passengers are secured in their seats. After the __9._____ the cabin attendants should help passengers to put on oxygen masks.

CONCLUSION

In emergency situations the crew must work quickly and calmly. The cabin crew must secure the cabin and take care of passenger safety. They should give instructions that are clear and easy to understand. The cockpit crew must confirm the aircraft condition and fly the airplane safely. Finally, if the captain cannot find a suitable __10._____, then he must prepare the crew for an emergency landing.

FLUENCY

CD48 CDで上の文を聞き、音読しましょう。何回か音読を繰り返します。

UNIT 9 — HOLDING

"THE AIRPORT IS NOW CLOSED"
BAD WEATHER & NATURAL DISASTERS

VOCABULARY

KEY WORDS

それぞれのキーワードにあった意味を選びましょう。例を参考にしてください。(回答は 86 ページ)

1. aircraft status — e — a. the period that an aircraft must wait for further instructions
2. fuel status — ___ — b. a round pattern that looks like an egg
3. hold time — ___ — c. the action of clearing a runway of obstructions
4. indefinitely — ___ — d. an electrical supply for use in emergencies
5. oval-shaped — ___ — e. the condition of a plane (eg: amount of fuel, damage)
6. racetrack — ___ — f. several airplanes flying in holding patterns above each other
7. re-open — ___ — g. a type of holding pattern for aircraft
8. runway sweep — ___ — h. for an unknown period of time
9. stack — ___ — i. to become operational again
10. standby power — ___ — j. the amount of fuel that an airplane has at the current time

PRONUNCIATION — CD49

CDで上の文を聞き、後についてリピートしましょう。何回か繰り返します。

PRONUNCIATION — CD50

CDで以下の表現を聞き、後についてリピートしましょう。

"L" IN VARIOUS POSITIONS

1. pilot
2. altitude
3. delay
4. controller
5. safety regulations
6. airport elevation
7. local time
8. flight level
9. light turbulence
10. landing clearance

To make the "L" sound: touch the tip of your tongue to the top of your mouth!

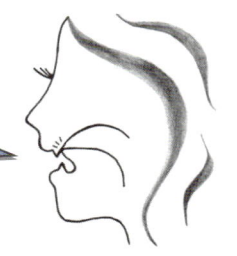

PRONUNCIATION — CD51

CDで以下の表現を聞き、後についてリピートしましょう。

"R" IN VARIOUS POSITIONS

1. crew
2. terrain
3. bearing
4. arrival
5. traffic separation
6. final approach
7. radio operator
8. priority landing
9. visual reference
10. full route clearance

To make the "R" sound: lift the middle of your tongue but don't touch the top of your mouth!

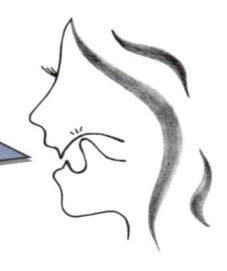

UNIT 9 — HOLDING

SINGLE PICTURE

FLUENCY

絵の状況について、SPRCモデルを使って、3分程度話しましょう。まず、左下の文を使って話し始めます。次に、右下のマインドマップの言葉を使って話すようにします。それをS、P、R、Cとそれぞれ行います。

*Organize your ideas with the **SPRC** model!*

S = SITUATION

This picture shows the situation at an airport…

- Above the airport
  - sky condition
  - holding patterns
  - how many aircraft?
- On the ground
  - how many runways?
  - any aircraft?
  - any vehicles?

P = PROBLEM

I think that maybe there was an earthquake…

- Earthquake
  - runway condition
  - what other damage?
  - standby power

Use present tenses for the single picture!

R = RESPONSE

It looks like this airport is closed at the moment…

- Airport closed
  - assess damage
  - runway sweep
  - when re-open?
- Holding aircraft
  - fuel status
  - landing order
  - continue hold or divert?

C = CONCLUSION

It is important to get accurate information as quickly as possible…

- Accurate information
  - airport staff actions
  - pilot actions
  - airport re-opening

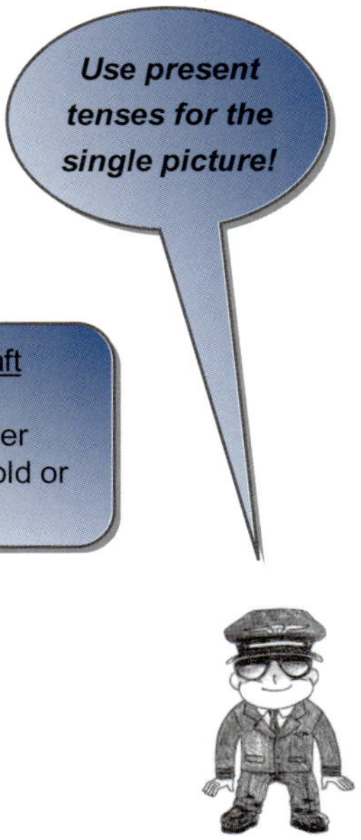

UNIT 9　　　　　　　　　　　　　　　　　　　　　　　　　　　　HOLDING

USEFUL LANGUAGE

STRUCTURE

このエクササイズでは、「want to」と「need to」の表現を使って、さまざまな人達がしたい事やする必要がある事を表わします。下の単語を並べ替えて文を作りましょう。1問目の例を参考にしてください。

WANTS & NEEDS

1. before landing. / to / the hold time / The pilots / know / want
   → _The pilots want to know the hold time before landing._
2. to / The controller / a runway / needs / order / sweep.
   → _____
3. need / fuel status. / monitor / The cockpit crews / their / to
   → _____
4. the flight. / to / wants / of / A hijacker / take control 
   → _____
5. needs / the hijacker's demands / to / to tell / the captain. / The CA
   → _____
6. The captain / to / to set / 7500. / the transponder / needs
   → _____
7. avoid / need / ash cloud. / to / The pilots / the volcanic 
   → _____
8. wants / a PIREP / the ash cloud. / to make / The captain / about
   → _____
9. want / the volcanic ash / The passengers / see / cloud. / to
   → _____
10. to find out / the problem. / need / of / the nature / The pilots 
    → _____
11. needs to / while / troubleshoots. / One pilot / fly the plane / the other
    → _____
12. wants / the problem. / the company / The captain / about / to contact
    → _____

PRONUNCIATION

CD52 CDで上の文を聞き、後についてリピートしましょう。何回か繰り返します。（回答は86ページ）

FLUENCY

上の4つの状況と同じような経験があれば、描写しながら話しましょう。

56

UNIT 9 — HOLDING

PICTURE SEQUENCE

Use past tenses for the picture sequence!

INTERACTIONS

CD53 1〜4の絵について以下の問題に答えましょう。まずは、CDでそれぞれの問題を聞きます。問題を聞いた後に一時停止ボタンを押して、それぞれの問題に対して回答します。完全文を使って、回答は声に出して言いましょう。

Picture 1
1. Describe the situation.
2. What was the problem?

Picture 2
3. What did the pilots see on final approach?
4. How did the pilots respond?

Picture 3
5. Where did the pilots go?
6. Describe the situation on the ground.

Picture 4
7. What other problem did the pilots have?
8. What do you think happened next?

UNIT 9 — HOLDING

VOCABULARY REVIEW

VOCABULARY

キーワードの復習です。ヒントを読んで、クロスワードに答えを書きましょう。(回答は 86 ページ)

ACROSS
5. for an unknown period of time
7. an electrical supply for use in emergencies
8. several airplanes flying in holding patterns above each other
9. a type of holding pattern for aircraft
10. the amount of fuel that an airplane has at the current time

DOWN
1. a round pattern that looks like an egg
2. the action of clearing a runway of obstructions
3. the condition of a plane (eg: amount of fuel, damage)
4. to become operational again
6. the period that an aircraft must wait for further instructions

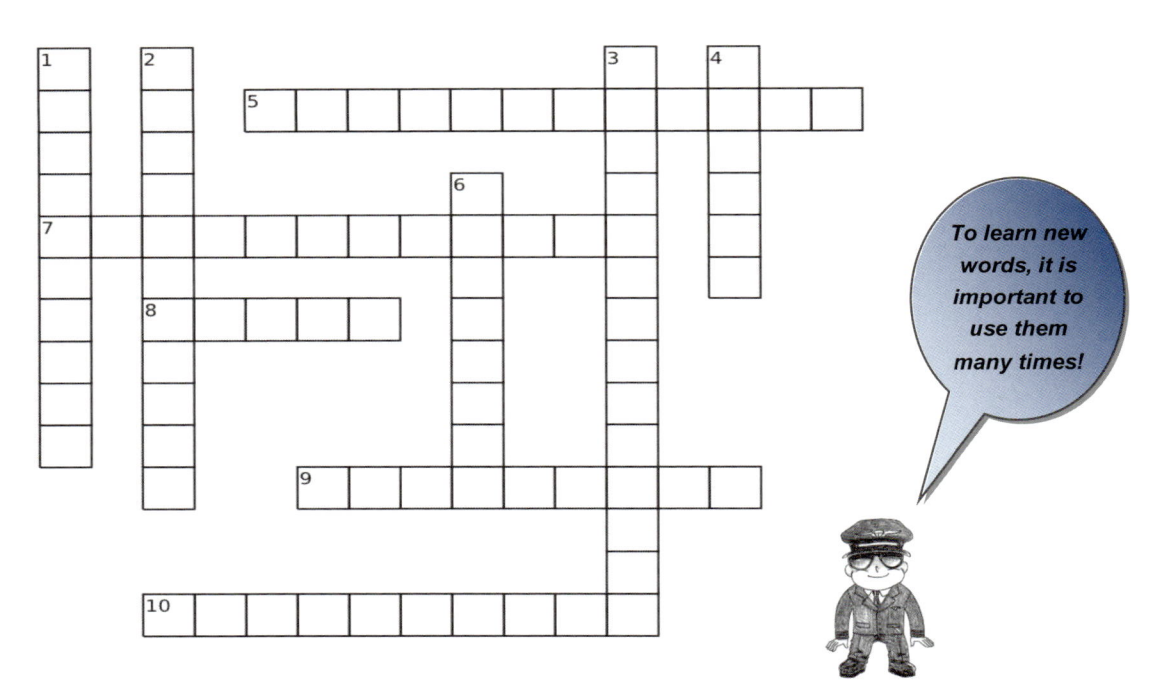

To learn new words, it is important to use them many times!

FLUENCY

上のいくつかのキーワードを使って、3つの文を作りましょう。その後文を何回か音読します。

SENTENCES

1. _____

2. _____

3. _____

UNIT 9

HOLDING

SINGLE PICTURE REVIEW

VOCABULARY

以下のキーワードを使って、下線部を埋めましょう。（回答は 86 ページ）

KEY WORDS

a. aircraft status
b. fuel status
c. hold time
d. indefinitely
e. oval-shaped
f. racetrack
g. re-open
h. runway sweep
i. stack
j. standby power

SITUATION

This picture shows the situation at an airport. The sky condition is scattered, and I can see three aircraft in __1._____ holding patterns above the airport. Each plane is flying an __2._____ course at a different altitude. ATC controllers instruct pilots to hold in a __3._____ when there is a delay at the airport. On the ground I can see a control tower and two runways, 34L and 34R. There is one aircraft on Runway 34R, and there are a number of ground vehicles on Runway 34L.

PROBLEM

I think that maybe there was an earthquake. I can see large cracks across Runway 34L, and some of the buildings seem to be shaking. Earthquakes can damage runways, taxiways and airport buildings. They can also affect the power supply. I think this airport is operating on __4._____ because I can see only a few lights in the control tower.

RESPONSE

It looks like this airport is closed at the moment. First the airport staff need to assess the damage. There are many vehicles and workers around Runway 34L so I think that a __5._____ is in progress now. If there is little damage, then the airport can re-open soon. However, the cracks seem to be large so I think that the airport might be closed __6._____. Meanwhile several aircraft are in holding patterns above the airport. The pilots of these planes have to check their __7._____ and make some decisions. In particular, they must check their __8._____ to see how long they can hold. In addition they need to check the airport condition and the landing order. The pilots may want to divert to alternate airports if the __9._____ is too long. However, other airports in the area might also have earthquake damage. Therefore it might be difficult to find a safe place to land.

CONCLUSION

It is important to get accurate information as quickly as possible. Airport staff need to assess the situation on the ground, and pilots need to check the status of their aircraft. Pilots must also consider the time it takes to __10._____ the airport. Even after re-opening, it may take a lot of time depending on the landing sequence of the airplanes.

FLUENCY

CD54 CDで上の文を聞き、音読しましょう。何回か音読を繰り返します。

UNIT 10 — APPROACH

"DOWN AND LOCKED?"
PROBLEMS DURING APPROACH

VOCABULARY

それぞれのキーワードにあった意味を選びましょう。例を参考にしてください。（回答は 87 ページ）

KEY WORDS

1. approach speed   c
2. ceiling   _____
3. final approach   _____
4. glide path   _____
5. glide slope   _____
6. noise abatement _____
7. residential area _____
8. runway incursion _____
9. runway number   _____
10. visibility   _____

a. a beam that provides vertical guidance on the final approach
b. a place where many people live
c. the airspeed that the pilot flies just before landing
d. an indication of the magnetic direction of an airstrip
e. a person or vehicle entering a runway without permission
f. the course followed by an aircraft as it descends to landing
g. the height of the base of a broken or overcast cloud layer
h. the greatest distance at which an object can be clearly seen
i. reducing the amount of noise around an airport
j. the last part of a flight when the pilot aligns a plane for landing

PRONUNCIATION

CD55 CDで上の文を聞き、後についてリピートしましょう。何回か繰り返します。

COMPREHENSION

CD56 CDで 5 つのATCの会話を聞き、下線部を埋めましょう。（回答は 87 ページ）

LISTENING TO ATC

1. PILOT: Control, Citation 1663, _____, FL 180. Request descent to a lower altitude due to _____.
2. CONTROLLER: All aircraft, RWY 16R now closed due to removal of _____ on the runway. Estimated _____ is 30 minutes.
3. PILOT: TWR, Citation 1663, we have a female passenger in her thirties with a head injury due to _____ encounter. Request _____ and ambulance upon landing.
4. CONTROLLER: King Air 1663, wind variable between _____, gusts 24, cleared to land RWY 16R. Caution, air speed plus minus 10 knots on _____ reported by a Boeing 737.
5. PILOT: TWR, Citation 1663, _____ over CHARLIE, we have a problem with our _____. Request lower altitude and weather at Akita Airport.

PRONUNCIATION

CD57 CDで上の文を聞き、後についてリピートしましょう。何回か繰り返します。

FLUENCY

上のそれぞれのATCの会話について、PROBLEMとRESPONSEを描写しながら話しましょう。

UNIT 10　　　　　　　　　　　　　　　　　　　　　　　　　　　　　　APPROACH

SINGLE PICTURE

FLUENCY

絵の状況について、SPRC モデルを使って、3 分程度話しましょう。まず、左下の文を使って話し始めます。次に、右下のマインドマップの言葉を使って話すようにします。それを S、P、R、C とそれぞれ行います。

*Organize your ideas with the **SPRC** model!*

Make sentences with the words in the mindmap!

S = SITUATION

I can see an airport with two parallel runways…

Airport area
- parallel runways
- where are the airplanes?
- residential area

P = PROBLEM

It looks like there are three potential problems…

Noise
- takeoffs & landings
- residential area

Weather
- visibility
- what kind of danger?

Runway incursion
- where is the plane?
- what kind of danger?

R = RESPONSE

There are several different responses to these problems…

Noise abatement
- engine thrust
- flap settings
- glide slope

Weather
- visibility, ceiling & RVR
- missed approach

Aircraft on runway
- go around
- contact controller

C = CONCLUSION

In conclusion, there are many things to consider when making an approach…

Approach conditions
- runway
- weather

Evaluate situation
- approach speed
- unstable approach
- safe to land?

UNIT 10 — USEFUL LANGUAGE

APPROACH

STRUCTURE

このエクササイズでは、「if」と「should」を使って、さまざまな状況に対するアドバイスを表わします。下の単語を並べ替えて文を作りましょう。1問目の例を参考にしてください。

CONDITIONAL ADVICE

1. high, / is / go around. / the pilot / If the approach / should
   → *If the approach is high, the pilot should go around.*

2. on final, / should not / If the plane / passengers / is / use cellphones.
   → _____

3. tower. / If the pilot / should contact / executes / he / a missed approach,
   → _____

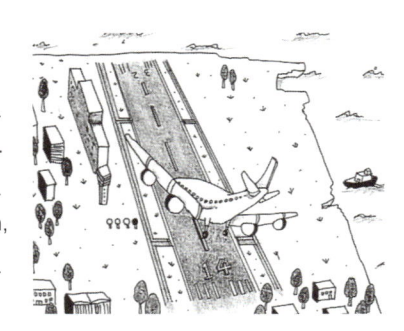

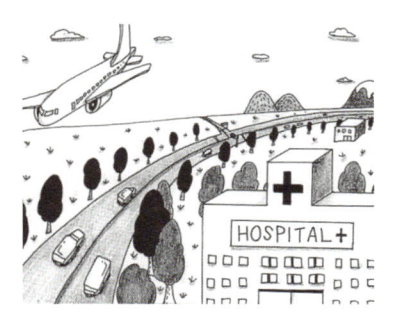

4. is / should / the pilot / in effect, / follow the procedures. / If noise abatement
   → _____

5. If the pilot / he / follow the procedures, / contact ATC. / cannot / should
   → _____

6. ATC / complains / should / If the hospital / about noise, / redirect traffic.
   → _____

7. a caution. / is / should issue / If the hot air balloon / the controller / too close,
   → _____

8. the pilot / If there are / go around. / on the runway, / should / obstructions
   → _____

9. change. / change, / If / the active runway / the wind conditions / should
   → _____

10. the pilot / If the approach / go around. / not stable, / should / is
    → _____

11. encounters / should / a microburst, / make a PIREP. / he / If the pilot
    → _____

12. continues, / should / If the microburst / the runway. / the controller / close
    → _____

PRONUNCIATION

CD58 CDで上の文を聞き、後についてリピートしましょう。何回か繰り返します。（回答は87ページ）

FLUENCY

上の4つの状況と同じような経験があれば、描写しながら話しましょう。

UNIT 10　　　　　　　　　　　　　　　　　　　　　　　　　　APPROACH

PICTURE SEQUENCE

Do not use Japanese words!

INTERACTIONS

CD59 1〜4の絵について以下の問題に答えましょう。まずは、CDでそれぞれの問題を聞きます。問題を聞いた後に一時停止ボタンを押して、それぞれの問題に対して回答します。完全文を使って、回答は声に出して言いましょう。

Picture 1
1. Describe the airplane.
2. How were the flight conditions?

Picture 2
3. What problem did the controller notice?
4. What were the pilot options?

Picture 3
5. What did the pilot say to the passengers?
6. How did the passengers respond?

Picture 4
7. Describe the pilot's approach to land.
8. What do you think happened next?

UNIT 10 — APPROACH

VOCABULARY REVIEW

VOCABULARY

キーワードの復習です。ヒントを読んで、クロスワードに答えを書きましょう。（回答は 87 ページ）

ACROSS

2. reducing the amount of noise around an airport
5. the height of the base of a broken or overcast cloud layer
7. a person or vehicle entering a runway without permission
9. the course followed by an aircraft as it descends to landing
10. a beam that provides vertical guidance on the final approach

DOWN

1. a place where many people live
3. the airspeed that the pilot flies just before landing
4. the greatest distance at which an object can be clearly seen
6. the last part of a flight when the pilot aligns a plane for landing
8. an indication of the magnetic direction of an airstrip

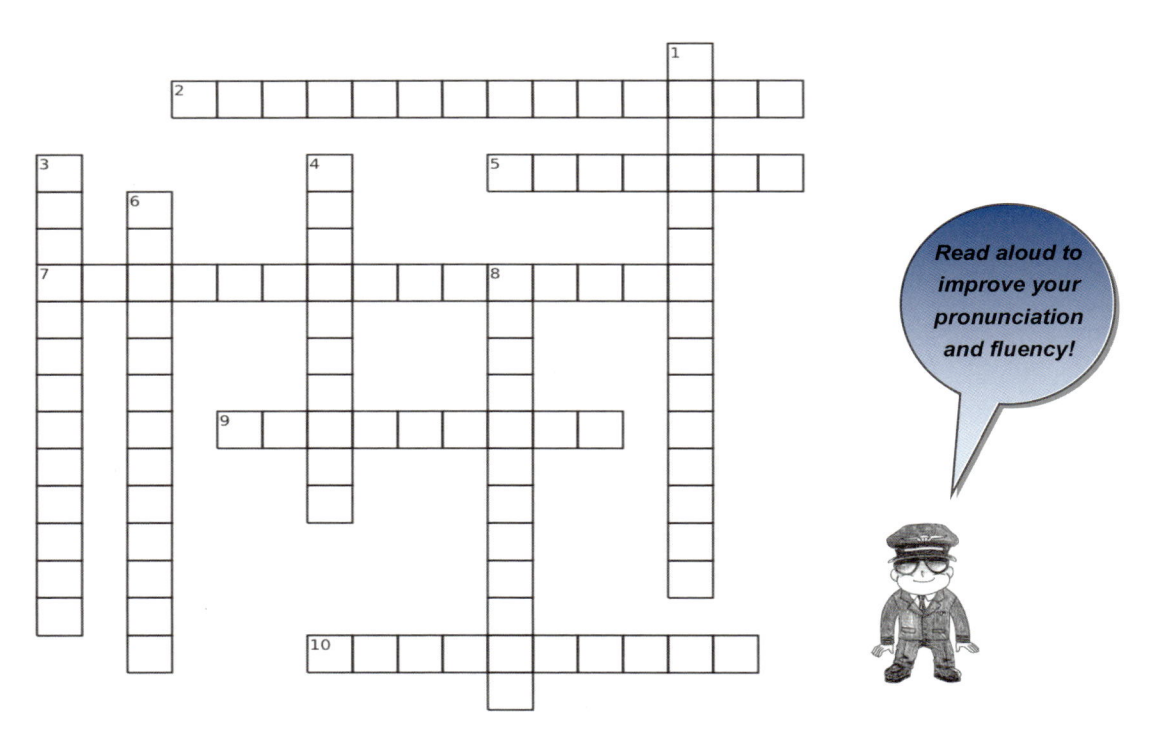

Read aloud to improve your pronunciation and fluency!

FLUENCY

上のいくつかのキーワードを使って、3 つの文を作りましょう。その後文を何回か音読します。

SENTENCES

1. _____

2. _____

3. _____

UNIT 10　　APPROACH

SINGLE PICTURE REVIEW

VOCABULARY
以下のキーワードを使って、下線部を埋めましょう。（回答は 87 ページ）

KEY WORDS

a. approach speed
b. ceiling
c. final approach
d. glide path
e. glide slope
f. noise abatement
g. residential area
h. runway incursion
i. runway numbers
j. visibility

SITUATION

I can see an airport with two parallel runways, but I cannot see the __1._____. One airplane is making a __2._____ to the airport, and another aircraft is taking off from the runway on the left. I can also see a hospital, some houses and some other buildings at the approach end of the runways, so I think that there is a __3._____ in the vicinity of this airport.

PROBLEM

It looks like there are three potential problems at this airport. The first problem is noise. Large jet aircraft make a lot of noise during takeoffs and landings. This can cause a lot of trouble for people who live and work in the area, especially if the __4._____ goes over a residential area. The second problem is the weather. The __5._____ seems to be poor in this picture. Therefore it might be difficult for the pilots to see the situation clearly. The final problem is a possible __6._____. It is difficult to see, but I think that an airplane is entering the runway on the right. If this plane enters the runway without clearance, it may create a problem.

RESPONSE

There are several different responses to these problems. Firstly, I think that __7._____ procedures may be in effect because there is a residential area near this airport. Therefore the pilots must be careful with engine thrust and flap settings. Also, they should not fly below the __8._____. Secondly, it looks like the visibility is poor and the __9._____ is low. If the RVR becomes too low, the pilots have to execute a missed approach. Finally, runway incursions are very dangerous. If another aircraft is on the runway, the pilots must not land. They should go around and contact the controller.

CONCLUSION

In conclusion, there are many things to consider when making an approach to an airport. The type of approach depends on the runway conditions, surrounding terrain and weather conditions. If the __10._____ is too fast, the aircraft might have a hard landing or an overrun. If the approach is unstable, the pilots should execute a missed approach. The pilots have to evaluate the situation and decide whether it is safe to land.

FLUENCY
CD60 CDで上の文を聞き、音読しましょう。何回か音読を繰り返します。

UNIT 11   LANDING

"GO AROUND, GO AROUND"
CROSSWINDS & WAKE TURBULENCE

VOCABULARY

それぞれのキーワードにあった意味を選びましょう。例を参考にしてください。（回答は 88 ページ）

KEY WORDS

1. clear the active   h
2. crabbing   _____
3. grass runway   _____
4. gust   _____
5. malfunction   _____
6. midfield   _____
7. obstacle   _____
8. tailwheel   _____
9. traffic flow   _____
10. wake turbulence   _____

a. movement of aircraft in the control area or airspace
b. an object that blocks the movement of an aircraft or airport vehicle
c. the middle part of the runway
d. a landing technique used for crosswind conditions
e. part of the landing gear located near the back of the fuselage
f. rough air that forms behind a plane as it passes through the air
g. a strong and sudden rush of wind
h. to move off the runway and onto a taxiway
i. a landing area that has a turf surface
j. a situation where airplane hardware is not working properly

PRONUNCIATION   CD61   CDで上の文を聞き、後についてリピートしましょう。何回か繰り返します。

COMPREHENSION   CD62   CDで5つのATCの会話を聞き、下線部を埋めましょう。（回答は 88 ページ）

LISTENING TO ATC

1. PILOT: TWR, Citation 1663, we are _____, RWY 18. There seems to be some kind of obstruction on the _____. It looks like an article of clothing.
2. CONTROLLER: Tower observation, fog bank developing _____ of the airport. Expect IMC conditions. For further information, monitor ATIS on _____.
3. PILOT: TWR, King Air 1663, we have a red light on our _____. Request low pass for _____.
4. CONTROLLER: Citation 1663, _____, RWY 18. Use caution, _____ reported towards the end of the runway.
5. PILOT: Controller, OB Air 1663, we made a go around due to _____. Request _____ and priority landing due to low fuel condition.

PRONUNCIATION   CD63   CDで上の文を聞き、後についてリピートしましょう。何回か繰り返します。

FLUENCY   上のそれぞれのATCの会話について、PROBLEMとRESPONSEを描写しながら話しましょう。

UNIT 11   LANDING

SINGLE PICTURE

FLUENCY

絵の状況について、SPRC モデルを使って、3 分程度話しましょう。まず、左下の文を使って話し始めます。次に、右下のマインドマップの言葉を使って話すようにします。それを S、P、R、C とそれぞれ行います。

*Organize your ideas with the **SPRC** model!*

S = SITUATION

In this picture I can see part of an airport…

- Airport
  - runways & taxiways
  - how many airplanes?
- Weather
  - visibility
  - wind conditions

⬇

P = PROBLEM

I think that there are several potential problems…

- Wind
  - windsock
  - gusts & crosswind
- Jet plane
  - wake turbulence
  - what kind of danger?
- Grass runway
  - truck
  - equipment malfunction

⬇

R = RESPONSE

There are several things that the pilots and controller should do…

- Crosswind
  - crabbing approach
  - wind check
  - go around
- Wake turbulence
  - issue caution
  - traffic flow
- Grass runway
  - tow truck away
  - clear the active

⬇

C = CONCLUSION

Landing is the most important phase of flight…

- Landing
  - important flight phase
  - pilot actions
  - controller actions

Use present tenses for the single picture!

UNIT 11   **LANDING**

USEFUL LANGUAGE

STRUCTURE

このエクササイズでは、「must」「has/have to」「because」を使って、とらなければならない行動とその理由を表わします。下の単語を並べ替えて文を作りましょう。1問目の例を参考にしてください。

NECESSARY ACTIONS & REASONS

1. must / snowing. / sweep the runways / it is / Airport staff / because
   → *Airport staff must sweep the runways because it is snowing.*

2. de-ice all aircraft / They / snow and ice. / there is / have to / because
   → _____

3. of / advise / The controller / all stations / the poor weather conditions. / must
   → _____

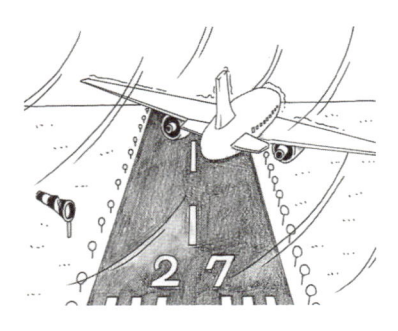

4. must / because / The pilot / there is / crab the plane / a strong crosswind.
   → _____

5. the approach / go around / The captain / because / is unstable. / has to
   → _____

6. due to / the active runway / must / the crosswind. / The controller / change
   → _____

7. must / smoke / shut down the engine / is coming out. / The pilots / because
   → _____

8. have to / due to / an emergency landing / They / the engine failure. / request
   → _____

9. prepare to / is on fire. / The crew / because / evacuate / the engine / must
   → _____

10. the weather / The captain / because / fly by instruments / is poor. / has to
    → _____

11. The passengers / is landing. / fasten seatbelts / the plane / because / must
    → _____

12. must / visibility. / the landing lights / Airport staff / poor / due to / turn on
    → _____

PRONUNCIATION

CD64 CDで上の文を聞き、後についてリピートしましょう。何回か繰り返します。(回答は88ページ)

FLUENCY

上の4つの状況と同じような経験があれば、描写しながら話しましょう。

UNIT 11   LANDING

PICTURE SEQUENCE

Use past tenses for the picture sequence!

INTERACTIONS **CD65** 1～4の絵について以下の問題に答えましょう。まずは、CDでそれぞれの問題を聞きます。問題を聞いた後に一時停止ボタンを押して、それぞれの問題に対して回答します。完全文を使って、回答は声に出して言いましょう。

Picture 1

1. What was the situation at this airport?
2. What were the airplanes doing?

Picture 2

3. How was the weather?
4. Describe the final approach.

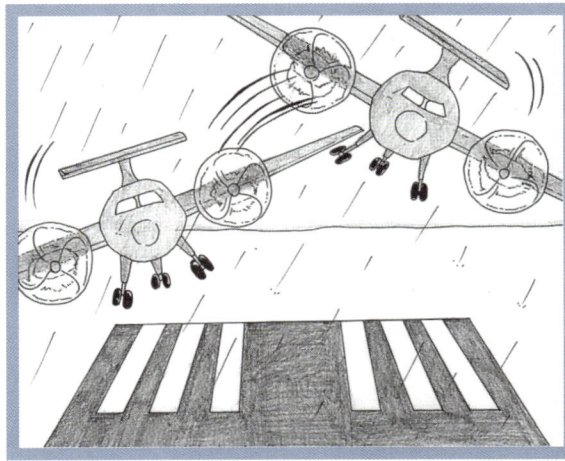

Picture 3

5. Why was the landing not stable?
6. What options did the pilot have?

Picture 4

7. What happened to the airplane?
8. What do you think happened next?

UNIT 11   LANDING

VOCABULARY REVIEW

VOCABULARY

キーワードの復習です。ヒントを読んで、クロスワードに答えを書きましょう。(回答は 88 ページ)

ACROSS

3. a situation where airplane hardware is not working properly
4. a landing technique used for crosswind conditions
7. a strong and sudden rush of wind
9. rough air that forms behind a plane as it passes through the air
10. an object that blocks the movement of an aircraft or airport vehicle

DOWN

1. movement of aircraft in the control area or airspace
2. to move off the runway and onto a taxiway
5. a landing area that has a turf surface
6. the middle part of the runway
8. part of the landing gear located near the back of the fuselage

To learn new words, it is important to use them many times!

FLUENCY

SENTENCES

上のいくつかのキーワードを使って、3つの文を作りましょう。その後文を何回か音読します。

1. _____
2. _____
3. _____

UNIT 11　　　　　　　　　　　　　　　　　　　　　　　　　　　　　　　　LANDING

SINGLE PICTURE REVIEW

VOCABULARY

以下のキーワードを使って、下線部を埋めましょう。(回答は 88 ページ)

KEY WORDS

a. clear the active
b. crabbing
c. grass runway
d. gusts
e. malfunction
f. midfield
g. obstacle
h. tailwheel
i. traffic flow
j. wake turbulence

SITUATION

In this picture I can see part of an airport with runways, taxiways and three airplanes. A large jet plane is taking off and a light aircraft is making an approach to land. It looks like the visibility is good, but there is a windsock at 1._____ and the wind seems to be very strong. There is another light aircraft with a tailwheel type of landing gear on a __2._____. This aircraft is next to a truck that is smoking.

PROBLEM

I think that there are several potential problems in this picture. First, the windsock indicates a strong crosswind. It is difficult for small planes to land if there are __3._____ of wind or a strong crosswind. Second, the jet plane is taking off and creating __4._____. This rough air can be dangerous for small aircraft, especially during takeoffs and landings. Finally, I think there is a problem with the __5._____ plane and the truck. It is difficult to see, but maybe it is some kind of equipment __6._____.

RESPONSE

There are several things that the pilots and controller should do in response to these problems. The pilot of the light aircraft is now making a turning approach. He has to use a __7._____ approach to adjust for the crosswind. The pilot should also request a wind check to confirm the wind conditions. If it is too difficult to land, the pilot must go around and contact the tower. If the wind conditions become too severe, the controller must change runways. Meanwhile, the controller has to order ground staff to tow the truck away because it is an __8._____ on the grass runway. Also, the controller should order the tailwheel aircraft to __9._____ runway. Finally, wake turbulence is dangerous for small planes so the controller should issue a caution to the arriving aircraft. He should also adjust the __10._____ to ensure safe separation between the departing and arriving traffic.

CONCLUSION

Landing is the most important phase of flight. Pilots must check the changing weather, airport and traffic conditions, and make adjustments in their approach. Controllers have to ensure the safe separation of traffic. They must also issue warnings and cautions so that aircraft can take off and land safely.

FLUENCY

CD66 CDで上の文を聞き、音読しましょう。何回か音読を繰り返します。

UNIT 12   AFTER LANDING

"REQUEST EMERGENCY ASSISTANCE"
OVERRUNS & OTHER MISHAPS

VOCABULARY
それぞれのキーワードにあった意味を選びましょう。例を参考にしてください。（回答は 89 ページ）

KEY WORDS

1. collision   i
2. deboard   ____
3. evacuation order   ____
4. first officer   ____
5. general aviation   ____
6. horizontal stabilizer   ____
7. skid off   ____
8. transmission   ____
9. widebody aircraft   ____
10. wingtip   ____

a. a large airliner with two or more aisles in the cabin
b. the pilot who is second in command of a flight
c. operating civilian planes for purposes other than commercial transport
d. to slide off the runway or taxiway
e. to get off an airplane
f. a command from the captain to quickly exit a plane in an emergency
g. the extreme outer edge of an aircraft wing
h. part of an aircraft tail section to which the elevators are attached
i. an accident that happens when a plane hits another object with force
j. a voice message sent by radio

PRONUNCIATION   CD67
CDで上の文を聞き、後についてリピートしましょう。何回か繰り返します。

COMPREHENSION   CD68
CDで 5 つのATCの会話を聞き、下線部を埋めましょう。（回答は 89 ページ）

LISTENING TO ATC

1. CONTROLLER: All stations, _____ reported during climb between _____ by Boeing 737, 10 minutes ago.
2. PILOT: Ground Control, King Air 1663, now _____. Unable to continue taxi due to _____. Request a tow back to south ramp.
3. CONTROLLER: Citation 1663, _____ on Taxiway D1. There will be a slight delay due to a King Air that needs a tow back to the _____.
4. PILOT: TWR, Citation 1663, _____, many birds on the approach path to RWY 18 and on the _____.
5. CONTROLLER: King Air 1663, your _____ are garbled and distorted. Cross RWY 18L and continue taxi to general aviation parking. Acknowledge by flashing your _____.

PRONUNCIATION   CD69
CDで上の文を聞き、後についてリピートしましょう。何回か繰り返します。

FLUENCY
上のそれぞれのATCの会話について、PROBLEMとRESPONSEを描写しながら話しましょう。

UNIT 12 AFTER LANDING

SINGLE PICTURE

FLUENCY

絵の状況について、SPRC モデルを使って、3 分程度話しましょう。まず、左下の文を使って話し始めます。次に、右下のマインドマップの言葉を使って話すようにします。それを S、P、R、C とそれぞれ行います。

*Organize your ideas with the **SPRC** model!*

Speak in complete sentences!

S = SITUATION

In this picture I can see an emergency situation…

Emergency situation
- skid off runway
- aircraft condition
- emergency vehicles

P = PROBLEM

I think that there are two serious problems…

Evacuation
- aircraft size
- how many passengers?

Fire risk
- engine smoke
- fuel leak

R = RESPONSE

The flight crew and the airport staff must respond quickly…

Flight crew
- evacuation order
- cabin crew actions

Emergency crew
- fire trucks
- put out fires
- medical assistance

Controller
- close airport
- transmission

C = CONCLUSION

In an emergency, the response time is very important…

Response time
- flight crew training
- emergency crew training
- coordination

UNIT 12   AFTER LANDING

USEFUL LANGUAGE

STRUCTURE

このエクササイズでは、「so」を使って、過去の状況に対する結果を表わします。下の単語を並べ替えて文を作りましょう。1問目の例を参考にしてください。

RESULTS IN THE PAST

1. was unruly / diverted to / A passenger / the pilot / so / another airport.
   → *A passenger was unruly so the pilot diverted to another airport.*
2. an emergency / a priority landing. / The pilot / he received / declared / so
   → _____
3. so / There was / called / the airport authorities / a problem / the police.
   → _____

4. taxied slowly. / extremely icy / the captain / The taxiways / so / were
   → _____
5. so / became / held his position. / worse / the captain / The ice and snow
   → _____
6. snowing / the airport. / It was / the controller / so / heavily / closed
   → _____

7. so / a collision. / there was / did not see / The pilots / the other aircraft
   → _____
8. were injured. / of the / It was / passengers / so / a heavy impact / some
   → _____
9. move / a long delay / The planes / there was / could not / so / at the airport.
   → _____

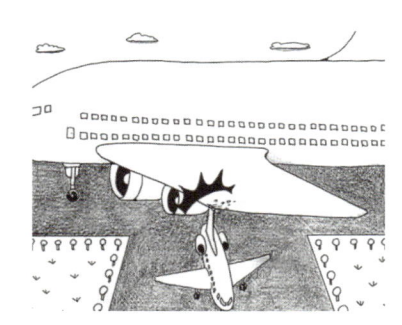

10. so / the boarding bridge. / was / hit / careless / The marshaller / the plane
    → _____
11. There was / surprised. / the passengers / a loud crashing sound / were / so
    → _____
12. damage / the next flight / to the fuselage / There was / so / was cancelled.
    → _____

PRONUNCIATION

CD70 CDで上の文を聞き、後についてリピートしましょう。何回か繰り返します。(回答は89ページ)

FLUENCY

上の4つの状況と同じような経験があれば、描写しながら話しましょう。

74

UNIT 12 — AFTER LANDING

PICTURE SEQUENCE

Keep your responses smooth and speak clearly!

INTERACTIONS

CD71 1〜4の絵について以下の問題に答えましょう。まずは、CDでそれぞれの問題を聞きます。問題を聞いた後に一時停止ボタンを押して、それぞれの問題に対して回答します。完全文を使って、回答は声に出して言いましょう。

Picture 1
1. What problem happened at this airport?
2. How did the airport staff respond to the problem?

Picture 2
3. Describe the evacuation procedure.
4. What problems happened during the evacuation?

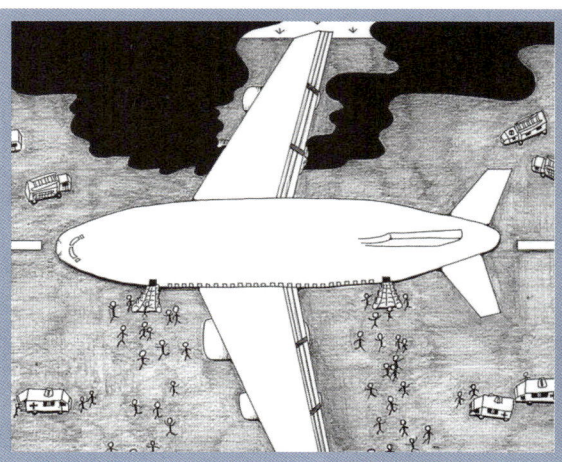

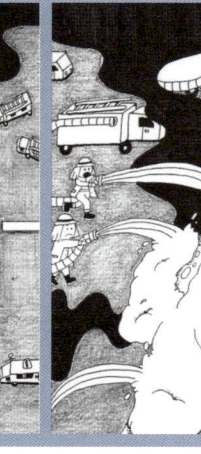

Picture 3
5. Which side of the plane did passengers exit from?
6. What did the passengers do after evacuating?

Picture 4
7. Describe the firefighting operation.
8. What do you think happened next?

UNIT 12 — AFTER LANDING

VOCABULARY REVIEW

VOCABULARY

キーワードの復習です。ヒントを読んで、クロスワードに答えを書きましょう。（回答は 89 ページ）

ACROSS

5. a command from the captain to quickly exit a plane in an emergency
6. to slide off the runway or taxiway
8. an accident that happens when a plane hits another object with force
9. a voice message sent by radio
10. a large airliner with two or more aisles in the cabin

DOWN

1. to get off an airplane
2. the pilot who is second in command of a flight
3. part of an aircraft tail section to which the elevators are attached
4. operating civilian planes for purposes other than commercial transport
7. the extreme outer edge of an aircraft wing

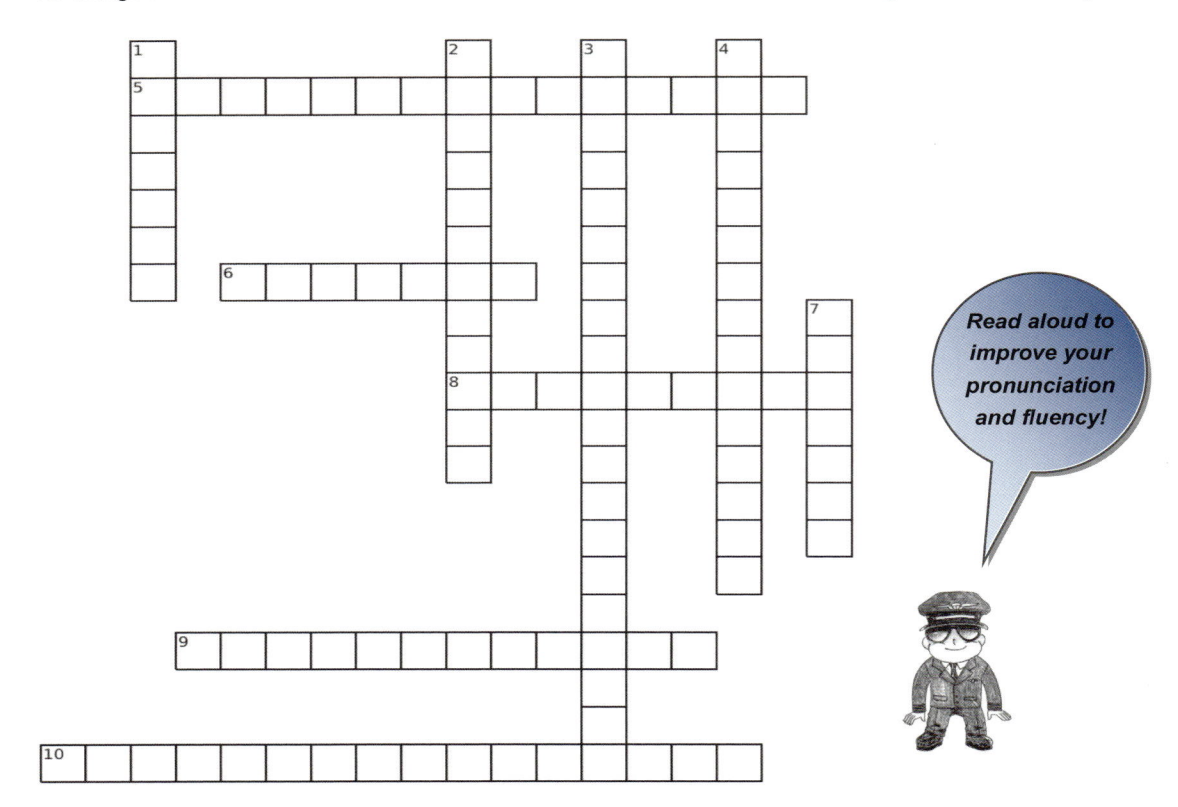

Read aloud to improve your pronunciation and fluency!

FLUENCY

上のいくつかのキーワードを使って、3 つの文を作りましょう。その後文を何回か音読します。

SENTENCES

1. _____
2. _____
3. _____

UNIT 12

AFTER LANDING

SINGLE PICTURE REVIEW

VOCABULARY

以下のキーワードを使って、下線部を埋めましょう。（回答は 89 ページ）

KEY WORDS

a. collision
b. deboard
c. evacuation order
d. first officer
e. general aviation
f. horizontal stabilizers
g. skidded off
h. transmission
i. widebody aircraft
j. wingtip

SITUATION

In this picture I can see an emergency situation. A large jet airliner has ___1._____ the end of a runway. The nose of the aircraft is covered with snow or soft earth, but the fuselage seems to be intact. I think that the vertical and ___2._____ are also intact. As a result of the ___3._____, the left engine and the left ___4._____ of the plane have broken off. A lot of smoke is coming from the engines, and one ambulance and two fire trucks are now approaching the plane.

PROBLEM

I think that there are two serious problems in this situation. The first problem is the evacuation. This plane is a ___5._____ so it might have a large number of passengers and crew. It can take a lot of time to evacuate the passengers, especially if some of them are injured. The second problem is the risk of fire. Thick smoke is coming from the engines, and there may be fuel leaks. The passengers and the crew must ___6._____ as quickly as possible because there is a danger of fire.

RESPONSE

The flight crew and the airport staff must respond quickly because this is an emergency situation. First, the captain has to give the ___7._____ to the cabin crew. Then the cabin attendants must help the passengers to exit the aircraft quickly and safely. Also, the captain or ___8._____ must shut down the fuel supply to the engines. Meanwhile, the fire trucks have to spray foam onto the engines to put out the fires. In addition, the emergency crew have to give medical assistance to injured passengers. Finally, the controller must close the airport and notify arriving aircraft about the emergency. After receiving this ___9._____, all commercial and ___10._____ traffic must divert to other airports nearby.

CONCLUSION

In an emergency, the response time is very important. A well-trained flight crew can work together to ensure that all the passengers evacuate safely and quickly. The emergency crew must put out the fires quickly and move the passengers and crew away from the crash site. Good coordination between the flight crew and airport staff is important for the success of the rescue operation.

FLUENCY

CD72 CDで上の文を聞き、音読しましょう。何回か音読を繰り返します。

UNIT 1 — PRE-FLIGHT OPERATIONS

ANSWERS

VOCABULARY — Page 6

KEY WORDS

1. h
2. a
3. j
4. b
5. i
6. c
7. d
8. f
9. e
10. g

STRUCTURE — Page 8, CD4

POSSIBLE PROBLEMS IN THE FUTURE

1. Heavy rain may reduce visibility for pilots.
2. Rain might make the runways slippery.
3. Slippery runways could cause problems during landing.
4. Heavy snow might cause flight delays.
5. Snow may cause icing on aircraft wings.
6. Icing on the wings could make the aircraft stall.
7. Strong winds may create turbulence.
8. Turbulence could cause injuries to passengers.
9. Pilots might lose control during severe turbulence.
10. A thunderstorm may produce a lot of thunder and lightning.
11. Loud thunder could scare some passengers.
12. Lightning might damage an aircraft's electrical systems.

Notice how some of the key words are used in these sentences!

VOCABULARY — Page 10

ACROSS
4. holding pattern
7. gate
8. runway
9. turbulence
10. microburst

DOWN
1. thunderstorm
2. adverse weather
3. hazard
5. lightning
6. delay

VOCABULARY — Page 11

KEY WORDS

1. i
2. h
3. d
4. f
5. j
6. g
7. a
8. c
9. b
10. e

UNIT 2 — AT THE RAMP

ANSWERS

VOCABULARY — Page 12

KEY WORDS

1. d
2. a
3. j
4. i
5. g
6. h
7. b
8. e
9. c
10. f

STRUCTURE — Page 14, CD10

EVENTS IN THE PAST

1. One aircraft and several vehicles were on the ramp.
2. Airport staff cleared the ramp of snow and ice.
3. Snow accumulated on the wings and fuselage of the plane.
4. The pilot went outside for the pre-flight inspection.
5. The pilot found a problem with a landing gear tire.
6. The pilot asked maintenance staff to fix the tire.
7. The ground crew loaded the cargo on to the aircraft.
8. A baggage handler loaded and secured the bags in the cargo hold.
9. The security dog checked the bags for dangerous goods.
10. A fuel truck and a baggage car were next to the plane.
11. The aircraft was loaded with fuel from the fuel truck.
12. The boarding stairs were removed from the aircraft's fuselage.

Repeat the sentences many times to improve your fluency!

VOCABULARY — Page 16

ACROSS

2. boarding stairs
5. ramp
7. winter operations
9. cancellation
10. cargo

DOWN

1. pre-flight inspection
3. fuselage
4. de-icing
6. precaution
8. taxiway

VOCABULARY — Page 17

KEY WORDS

1. h
2. e
3. a
4. b
5. i
6. j
7. d
8. c
9. g
10. f

UNIT 3　　　　　　　　　　　　　　　　　　　　　　　　GROUND MOVEMENT

ANSWERS

VOCABULARY — Page 18

KEY WORDS
1. b
2. h
3. f
4. c
5. g
6. j
7. d
8. a
9. e
10. i

STRUCTURE — Page 20, CD16

DESCRIBING OBSTRUCTIONS

1. There is some kind of obstruction in front of the airplane.
2. It looks like a bottle and some boxes are on the taxiway.
3. The pilot will contact the controller to remove the obstruction.
4. There are many vehicles in the ramp area.
5. It looks like something fell from the baggage car.
6. I think that the driver of the baggage car noticed the problem.
7. There seems to be an obstruction on the taxiway.
8. It looks like some work gloves and clothes.
9. The pilot stopped the airplane because of the obstruction.
10. It looks like something fell from the airplane.
11. It seems to be a panel from the fuselage.
12. The controller will contact the pilot about the missing panel.

Be careful with "L" and "R" sounds when saying these key words!

VOCABULARY — Page 22

ACROSS
2. obstruction
7. pushback
8. bagggage
9. baggage car
10. hydraulic fluid

DOWN
1. hold your position
3. taxi
4. slippery
5. braking action
6. leak

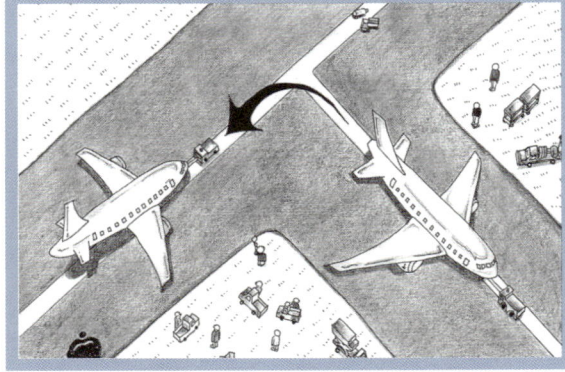

VOCABULARY — Page 23

KEY WORDS
1. b
2. a
3. h
4. g
5. f
6. c
7. d
8. e
9. i
10. j

Answer Key

80

UNIT 4 — CLEARED FOR TAKEOFF

ANSWERS

VOCABULARY — Page 24

KEY WORDS

1. c
2. f
3. i
4. e
5. h
6. b
7. j
8. d
9. a
10. g

COMPREHENSION — Page 24

LISTENING TO ATC

1. hold your position
   Delta 2
2. RWY 18R
   bird sweep
3. rejected takeoff
   tow assistance
4. braking action
   snow and ice
5. 15-minute delay
   baggage

STRUCTURE — Page 26, CD22

NECESSARY ACTIONS

1. There is some kind of obstruction on the runway.
2. The pilots have to reject takeoff due to the obstruction.
3. The pilots must contact the tower and report the incident.
4. There are two airplanes on the active runway.
5. The small airplane has to clear the runway.
6. The pilot must reject takeoff and hold his position.
7. There seems to be fluid leaking from the fuselage.
8. The pilots have to check the aircraft systems to find the problem.
9. The pilots must contact the controller and describe the situation.
10. It looks like this aircraft has a landing gear problem.
11. The controller must direct air traffic to go around.
12. The pilot has to request tow assistance to get off the runway.

VOCABULARY — Page 28

ACROSS

1. landing gear
4. bird strike
5. fire equipment
8. re-schedule
10. tow assistance

DOWN

2. rejected takeoff
3. incident
6. go around
7. bird sweep
9. stand by

VOCABULARY — Page 29

KEY WORDS

1. i
2. f
3. e
4. a
5. g
6. j
7. c
8. d
9. h
10. b

UNIT 5 — TAKEOFF & CLIMB

ANSWERS

VOCABULARY — Page 30

KEY WORDS

1. e
2. h
3. j
4. f
5. b
6. i
7. c
8. d
9. a
10. g

COMPREHENSION — Page 30

LISTENING TO ATC

1. caution
   departing 737
2. stuck landing gear
   air turnback
3. 30 minutes
   nose gear
4. microburst
   160 at 30
5. slippery
   ramp area

STRUCTURE — Page 32, CD28

EVENTS HAPPENING NOW

1. The airplane is climbing away from the airport.
2. The captain is trying to retract the flaps.
3. The captain must return to the airport and fix the problem.
4. The weather is very cold and snow is falling.
5. Ice is building up on the wings and fuselage.
6. The captain must fly out of the icing conditions.
7. The aircraft is experiencing an equipment failure.
8. A panel is falling from one of the engines.
9. The captain should contact ATC and request an air turnback.
10. One passenger is holding his chest and looks sick.
11. A cabin attendant is notifying the captain.
12. The sick passenger must receive immediate medical attention.

VOCABULARY — Page 34

ACROSS

3. equipment failure
4. EICAS
5. nose gear
8. injured passenger
10. stuck landing gear

DOWN

1. retract position
2. baggage compartment
6. air turnback
7. caution
9. flap

VOCABULARY — Page 35

KEY WORDS

1. c
2. e
3. h
4. g
5. i
6. j
7. a
8. b
9. d
10. f

UNIT 6 — CLIMB

ANSWERS

VOCABULARY — Page 36

KEY WORDS
1. g
2. d
3. i
4. f
5. h
6. a
7. c
8. j
9. b
10. e

COMPREHENSION — Page 36

LISTENING TO ATC
1. PIREP
   ash cloud
2. air turnback
   sick passenger
3. rough air
   FL150 and FL170
4. stuck landing gear
   nose gear
5. light to moderate turbulence
   15 miles northwest

STRUCTURE — Page 38, CD34

INTERRUPTED ACTIONS IN THE PAST
1. The plane was entering clouds when it encountered turbulence.
2. The pilot was levelling off when a passenger was injured.
3. One of the CAs notified the captain about the injured passenger.
4. A Cessna was changing heading when the engine overheated.
5. The pilot was looking for a landing area when the engine stopped.
6. The pilot declared an emergency and landed on a golf course.
7. The plane was entering clouds when it encountered icing conditions.
8. Ice was building up on the wings when the crew made a visual check.
9. The captain had to activate the aircraft's anti-icing systems.
10. The plane was climbing when a baggage compartment opened.
11. A passenger was talking to his wife when he was hit by a bag.
12. Cabin crew had to give medical assistance to the passenger.

VOCABULARY — Page 40

ACROSS
5. volcanic eruption
6. malfunction
7. ash cloud
9. terrain
10. landing area

DOWN
1. non-normal situation
2. PIREP
3. windshield
4. visual check
8. level off

VOCABULARY — Page 41

KEY WORDS
1. f
2. c
3. e
4. h
5. i
6. d
7. a
8. j
9. b
10. g

UNIT 7 — CRUISE

ANSWERS

VOCABULARY — Page 42

KEY WORDS

1. i
2. e
3. g
4. a
5. h
6. d
7. j
8. f
9. c
10. b

STRUCTURE — Page 44, CD40

SOLVING PROBLEMS

1. The boy should tell the CA exactly where it hurts.
2. The chief purser should inform the pilots about the sick passenger.
3. The captain should make a doctor call for medical assistance.
4. The husband should explain the situation to the CA.
5. The CA should move the pregnant passenger to a comfortable seat.
6. The captain should contact the company for advice.
7. The passenger should not play loud music on his computer.
8. The CA should politely ask the boy to turn down the volume.
9. The boy should put away his computer before takeoff and landing.
10. The medics should quickly take the sick passenger to hospital.
11. The captain should tell the other passengers to stay calm.
12. The other passengers should not leave their seats yet.

Be careful with "B" and "V" sounds when saying these key words!

VOCABULARY — Page 46

ACROSS

3. destination
7. police assistance
8. restrain
9. drunk
10. diversion

DOWN

1. medical assistance
2. unruly behavior
4. chief purser
5. evacuation
6. divert

VOCABULARY — Page 47

KEY WORDS

1. a
2. e
3. g
4. f
5. j
6. i
7. d
8. b
9. h
10. c

UNIT 8   EMERGENCY

ANSWERS

VOCABULARY — Page 48

KEY WORDS
1. h
2. f
3. a
4. g
5. j
6. d
7. b
8. i
9. c
10. e

STRUCTURE — Page 50, CD46

ACTIONS IN THE FUTURE

1. The cabin attendant is going to get a fire extinguisher.
2. She will use the fire extinguisher to put out the fire.
3. The chief purser will inform the captain about the emergency.
4. The CA is going to ask the passenger to put away the bottle.
5. Other passengers are going to complain about the drunk man.
6. The captain will ask the company for advice about the passenger.
7. All of the passengers are going to put on oxygen masks.
8. The pilot will contact ATC and request an emergency landing.
9. The CAs are going to prepare for an emergency evacuation.
10. All of the passengers are going to evacuate the airplane.
11. The CAs will assist old passengers and children to leave.
12. Emergency staff will help injured passengers outside the plane.

Repeat the sentences many times to improve your fluency!

VOCABULARY — Page 52

ACROSS
2. cruise flight
4. alternate airport
5. oxygen mask
7. radar vectors
8. rapid decompression
9. priority landing

DOWN
1. emergency procedure
2. controlled descent
3. flight plan
6. warning

VOCABULARY — Page 53

KEY WORDS
1. c
2. f
3. i
4. d
5. j
6. e
7. g
8. h
9. b
10. a

UNIT 9 — HOLDING

ANSWERS

VOCABULARY — Page 54

KEY WORDS
1. e
2. j
3. a
4. h
5. b
6. g
7. i
8. c
9. f
10. d

STRUCTURE — Page 56, CD52

WANTS & NEEDS

1. The pilots want to know the hold time before landing.
2. The controller needs to order a runway sweep.
3. The cockpit crews need to monitor their fuel status.
4. A hijacker wants to take control of the flight.
5. The CA needs to tell the hijacker's demands to the captain.
6. The captain needs to set the transponder to 7500.
7. The pilots need to avoid the volcanic ash cloud.
8. The captain wants to make a PIREP about the ash cloud.
9. The passengers want to see the volcanic ash cloud.
10. The pilots need to find out the nature of the problem.
11. One pilot needs to fly the plane while the other troubleshoots.
12. The captain wants to contact the company about the problem.

Try to use sentences like these when you describe pictures!

VOCABULARY — Page 58

ACROSS
5. indefinitely
7. standby power
8. stack
9. racetrack
10. fuel status

DOWN
1. oval-shaped
2. runway sweep
3. aircraft status
4. re-open
6. hold time

VOCABULARY — Page 59

KEY WORDS
1. f
2. e
3. i
4. j
5. h
6. d
7. a
8. b
9. c
10. g

UNIT 10 — APPROACH

ANSWERS

VOCABULARY — Page 60

KEY WORDS
1. c
2. g
3. j
4. f
5. a
6. i
7. b
8. e
9. d
10. h

COMPREHENSION — Page 60

LISTENING TO ATC

1. over DELTA
   turbulence
2. obstruction
   time of re-opening
3. moderate turbulence
   priority landing
4. 180 to 220 at 18
   final approach
5. FL150
   de-icing system

STRUCTURE — Page 62, CD58

CONDITIONAL ADVICE

1. If the approach is high, the pilot should go around.
2. If the plane is on final, passengers should not use cellphones.
3. If the pilot executes a missed approach, he should contact tower.
4. If noise abatement is in effect, the pilot should follow the procedures.
5. If the pilot cannot follow the procedures, he should contact ATC.
6. If the hospital complains about noise, ATC should redirect traffic.
7. If the hot air balloon is too close, the controller should issue a caution.
8. If there are obstructions on the runway, the pilot should go around.
9. If the wind conditions change, the active runway should change.
10. If the approach is not stable, the pilot should go around.
11. If the pilot encounters a microburst, he should make a PIREP.
12. If the microburst continues, the controller should close the runway.

VOCABULARY — Page 64

ACROSS
2. noise abatement
5. ceiling
7. runway incursion
9. glide path
10. glide slope

DOWN
1. residential area
3. approach speed
4. visibility
6. final approach
8. runway number

VOCABULARY — Page 65

KEY WORDS
1. i
2. c
3. g
4. d
5. j
6. h
7. f
8. e
9. b
10. a

UNIT 11 — LANDING

ANSWERS

VOCABULARY — Page 66

KEY WORDS
1. h
2. d
3. i
4. g
5. j
6. c
7. b
8. e
9. a
10. f

COMPREHENSION — Page 66

LISTENING TO ATC
1. going around
   runway, about midfield
2. 3 miles north
   124.05
3. nose gear
   visual check
4. cleared to land
   oil and rubber
5. wind shear
   radar vectors

STRUCTURE — Page 68, CD64

NECESSARY ACTIONS & REASONS
1. Airport staff must sweep the runways because it is snowing.
2. They have to de-ice all aircraft because there is snow and ice.
3. The controller must advise all stations of the poor weather conditions.
4. The pilot must crab the plane because there is a strong crosswind.
5. The captain has to go around because the approach is unstable.
6. The controller must change the active runway due to the crosswind.
7. The pilots must shut down the engine because smoke is coming out.
8. They have to request an emergency landing due to the engine failure.
9. The crew must prepare to evacuate because the engine is on fire.
10. The captain has to fly by instruments because the weather is poor.
11. The passengers must fasten seatbelts because the plane is landing.
12. Airport staff must turn on the landing lights due to poor visibility.

VOCABULARY — Page 70

ACROSS
3. malfunction
4. crabbing
7. gust
9. wake turbulence
10. obstacle

DOWN
1. traffic flow
2. clear the active
5. grass runway
6. midfield
8. tailwheel

VOCABULARY — Page 71

KEY WORDS
1. f
2. c
3. d
4. j
5. h
6. e
7. b
8. g
9. a
10. i

UNIT 12 — AFTER LANDING

ANSWERS

VOCABULARY — Page 72
KEY WORDS
1. i
2. e
3. f
4. b
5. c
6. h
7. d
8. j
9. a
10. g

COMPREHENSION — Page 72
LISTENING TO ATC
1. moderate turbulence
   FL250 and FL300
2. holding on Alpha 4
   poor visibility
3. hold your position
   ramp area
4. pilot report
   runway threshold
5. transmissions
   landing lights

STRUCTURE — Page 74, CD70
RESULTS IN THE PAST
1. A passenger was unruly so the pilot diverted to another airport.
2. The pilot declared an emergency so he received a priority landing.
3. There was a problem so the airport authorities called the police.
4. The taxiways were extremely icy so the captain taxied slowly.
5. The ice and snow became worse so the captain held his position.
6. It was snowing heavily so the controller closed the airport.
7. The pilots did not see the other aircraft so there was a collision.
8. It was a heavy impact so some of the passengers were injured.
9. The planes could not move so there was a long delay at the airport.
10. The marshaller was careless so the plane hit the boarding bridge.
11. There was a loud crashing sound so the passengers were surprised.
12. There was damage to the fuselage so the next flight was cancelled.

VOCABULARY — Page 76
ACROSS
5. evacuation order
6. skid off
8. collision
9. transmission
10. widebody aircraft

DOWN
1. deboard
2. first officer
3. horizontal stabilizer
4. general aviaiton
7. wingtip

VOCABULARY — Page 77
KEY WORDS
1. g
2. f
3. a
4. j
5. i
6. b
7. c
8. d
9. h
10. e

SELF-EVALUATION

This page helps check your English proficiency in the 6 areas of evaluation. Read each sentence below and circle "True" or "False".

After you have finished, check the sentences circled "False". Which areas of evaluation are they? Practice these areas more.

To achieve Level 4 on the English language proficiency test, you should be able to answer "True" for all the sentences below. Do your best!

What are your weak areas?

PRONUNCIATION

- My pronunciation is clear and accurate. TRUE FALSE
- My pronunciation is usually easy to understand. TRUE FALSE

STRUCTURE

- I speak in complete sentences. TRUE FALSE
- I control my structure and tenses. TRUE FALSE
- I do not make mistakes that change the meaning of my sentences. TRUE FALSE

VOCABULARY

- I know a wide range of aviation vocabulary. TRUE FALSE
- I make accurate word choices when needed. TRUE FALSE

FLUENCY

- I can speak at length on aviation topics. TRUE FALSE
- My speech is smooth and is not broken by fillers or pauses. TRUE FALSE
- I connect my ideas with discourse markers (eg: "so", "because", "therefore"). TRUE FALSE

COMPREHENSION

- I usually understand what other speakers are saying. TRUE FALSE
- I always check and confirm misunderstandings. TRUE FALSE

INTERACTIONS

- I can maintain communication exchanges. TRUE FALSE
- I offer enough information. TRUE FALSE
- My responses are timely and accurate. TRUE FALSE

Simon Cookson is an Associate Professor in the Aviation Management Department at J. F. Oberlin University. He has a Ph. D. in Sociolinguistics from International Christian University, a Master's degree in Aerospace Systems Engineering from the University of Southampton, and a Master's degree in Teaching English to Speakers of Other Languages (TESOL) from Aston University.

サイモン・クックソン

桜美林大学アビエーションマネジメント学類准教授。国際基督教大学社会言語学博士号、英国サウサンプトン大学院、航空宇宙システムエンジニアリング修士号、および英国アストン大学院、英語教授法修士号を持つ。

Michael Kelly worked for 23 years in the Flight Training Department at Japan Airlines managing the aviation and ATC English programs for pilot trainees. He is now an Assistant Professor in the Aviation Management Department at J. F. Oberlin University.

マイケル・ケリー

23年間、日本航空のフライトトレーニングデパートメント勤務。パイロット訓練生の航空英語とATC英語のプログラム管理を行った。現在は桜美林大学アビエーションマネジメント学類専任講師。

パイロットのための
ICAO航空英語能力試験ワークブック　定価はカバーに表示してあります

2013年 9 月 8 日　初版発行
2021年12月 8 日　3版発行

著　者　サイモン・クックソン，マイケル・ケリー
発行者　小川　典子
印　刷　倉敷印刷株式会社
製　本　東京美術紙工協業組合

発行所　株式会社　成山堂書店

〒160-0012　東京都新宿区南元町4番51 成山堂ビル
TEL：03(3357)5861　　FAX：03(3357)5867
URL　http://www.seizando.co.jp
落丁・乱丁本はお取り換えいたしますので、小社営業チーム宛にお送りください。

Ⓒ 2013　Simon Cookson, Michael Kelly
Printed in Japan　　　　ISBN978-4-425-86231-3

音声ファイルのダウンロードとご利用について

■音声ファイルは、下記のサイトからダウンロードしてご利用ください。全部で 72 ファイル（約 2 時間 30 分）あります。なお、音声ファイルは無料です。
http://www.seizando.co.jp/icao

■本文中の〔CD1〕〔CD2〕〔CD3〕……の表記は、上記の音声ファイル名〔Unit1 CD001〕〔Unit1 CD002〕〔Unit1 CD003〕……に対応しています。

■ファイル形式は mp3 です。対応する機器（PC、デジタルオーディオプレーヤー等）でご利用ください。音楽 CD 専用機をご利用の場合は、mp3 ファイルをオーディオファイルに変換するなどしてご利用ください。

■音声データの複製は個人で利用する範囲とします。著者または出版社の許可なく、他者に複製を供与する、またはインターネット上に公開する等の無断複製・頒布行為を禁じます。